Moving From R&D Development to Production and Sales in Start-Up Companies

Amiram Porath
APC, Israel

A volume in the Advances in Logistics, Operations, and Management Science (ALOMS) Book Series

IGI Global
PUBLISHER of TIMELY KNOWLEDGE

Published in the United States of America by
IGI Global
Business Science Reference (an imprint of IGI Global)
701 E. Chocolate Avenue
Hershey PA, USA 17033
Tel: 717-533-8845
Fax: 717-533-8661
E-mail: cust@igi-global.com
Web site: http://www.igi-global.com

Library of Congress Cataloging-in-Publication Data

Names: Porath, Amiram, author.
Title: Moving from R&D development to production and sales in start-up companies / by Amiram Porath.
Description: Hershey, PA : Business Science Reference, [2023] | Includes bibliographical references and index. | Summary: "This book focuses on the characteristics of start-ups that differentiate them from other companies"-- Provided by publisher.
Identifiers: LCCN 2020012897 (print) | LCCN 2020012898 (ebook) | ISBN 9781799856856 (hardcover) | ISBN 9781799856863 (paperback) | ISBN 9781799856870 (ebook)
Subjects: LCSH: Research, Industrial. | Technological innovations. | New products. | New business enterprises. | High technology industries.
Classification: LCC T175 .P6185 2021 (print) | LCC T175 (ebook) | DDC 658.1/1--dc23
LC record available at https://lccn.loc.gov/2020012897
LC ebook record available at https://lccn.loc.gov/2020012898

This book is published in the IGI Global book series Advances in Logistics, Operations, and Management Science (ALOMS) (ISSN: 2327-350X; eISSN: 2327-3518)

British Cataloguing in Publication Data
A Cataloguing in Publication record for this book is available from the British Library.

All work contributed to this book is new, previously-unpublished material.
The views expressed in this book are those of the authors, but not necessarily of the publisher.

For electronic access to this publication, please contact: eresources@igi-global.com.

Advances in Logistics, Operations, and Management Science (ALOMS) Book Series

John Wang
Montclair State University, USA

ISSN:2327-350X
EISSN:2327-3518

MISSION

Operations research and management science continue to influence business processes, administration, and management information systems, particularly in covering the application methods for decision-making processes. New case studies and applications on management science, operations management, social sciences, and other behavioral sciences have been incorporated into business and organizations real-world objectives.

The **Advances in Logistics, Operations, and Management Science** (ALOMS) Book Series provides a collection of reference publications on the current trends, applications, theories, and practices in the management science field. Providing relevant and current research, this series and its individual publications would be useful for academics, researchers, scholars, and practitioners interested in improving decision making models and business functions.

COVERAGE

- Risk Management
- Decision analysis and decision support
- Finance
- Networks
- Political Science
- Production Management
- Operations Management
- Marketing engineering
- Services management
- Organizational Behavior

IGI Global is currently accepting manuscripts for publication within this series. To submit a proposal for a volume in this series, please contact our Acquisition Editors at Acquisitions@igi-global.com or visit: http://www.igi-global.com/publish/.

The Advances in Logistics, Operations, and Management Science (ALOMS) Book Series (ISSN 2327-350X) is published by IGI Global, 701 E. Chocolate Avenue, Hershey, PA 17033-1240, USA, www.igi-global.com. This series is composed of titles available for purchase individually; each title is edited to be contextually exclusive from any other title within the series. For pricing and ordering information please visit http://www.igi-global.com/book-series/advances-logistics-operations-management-science/37170. Postmaster: Send all address changes to above address.

Titles in this Series

For a list of additional titles in this series, please visit: http://www.igi-global.com/book-series/

Cases on International Business Logistics in the Middle East
Islam Abdelbary (University of Plymouth, UK) Sandra Haddad (Arab Academy for Science, Technology, and Maritime Transport, Egypt) and Ghada Elkady (Arab Academy for Science, Technology, and Maritime Transport, Egypt)
Business Science Reference • © 2023 • 325pp • H/C (ISBN: 9781668446867) • US $240.00

Perspectives on Women in Management and the Global Labor Market
Elisabete S. Vieira (GOVCOPP, University of Aveiro, Portugal) Mara Madaleno (GOVCOPP, University of Aveiro, Portugal) and João Teodósio (Polytechnic Institute of Santarém, Portugal)
Business Science Reference • © 2023 • 335pp • H/C (ISBN: 9781668459812) • US $240.00

Cases on Managing Dairy Productive Chains
Eduardo Guilherme Satolo (São Paulo State University, Brazil) and Priscilla Ayllen Bustos Mac-Lean (São Paulo State University (UNESP), Brazil)
Engineering Science Reference • © 2023 • 310pp • H/C (ISBN: 9781668454725) • US $235.00

Examining the Vital Financial Role of SMEs in Achieving the Sustainable Development Goals
Dalila Taleb (University of Abou Bakr Belkaïd, Algeria) Mohammed El Amine Abdelli (University of Brest, France) Afef Khalil (University of Tunis, Carthage, Tunisia) and Asma Sghaier (University of Sousse, Tunisia)
Business Science Reference • © 2023 • 275pp • H/C (ISBN: 9781668448342) • US $250.00

Lean Thinking in Industry 4.0 and Services for Society
Katarzyna Antosz (Rzeszow University of Technology, Poland) Jose Carlos Sa (School of Engineering, ISEP, Polytechnic of Porto, Portugal) Małgorzata Jasiulewicz-Kaczmarek (Poznan University of Technology, Poland) and Jose Machado (University of Minho, Portugal)
Business Science Reference • © 2023 • 297pp • H/C (ISBN: 9781668456064) • US $250.00

IGI Global
PUBLISHER of TIMELY KNOWLEDGE

701 East Chocolate Avenue, Hershey, PA 17033, USA
Tel: 717-533-8845 x100 • Fax: 717-533-8661
E-Mail: cust@igi-global.com • www.igi-global.com

Table of Contents

Preface vii

Introduction x

Chapter 1
Literature Review 1

Chapter 2
The R&D-Based Start-Up 13

Chapter 3
Other New Firm Types 37

Chapter 4
The "Complete" Firm 53

Chapter 5
The Nature of the Gap 70

Chapter 6
Overcoming the Gap 85

Chapter 7
Government Tools and Assistive Tools 97

Chapter 8
SeeTRee Ltd. A Case Study 116

Chapter 9
The Learning Process as a Maturing Process for TSUs 124

Conclusion 133

Glossary 135

Compilation of References 139

Related References 144

About the Author 179

Index 180

Preface

When I started working in the innovation sector (not know that it would be such referred to in the future) in the early 1990's, the obvious way for a new start-up was to look towards an IPO. Not that there were no deals that ended in an M&A by a large firm, but most entrepreneurs wanted to build their own empire and the time to IPO was an important measure of how fast you wanted to grow. The M&A was used by large companies as a means to reach their growth targets. When at the end of the year the CEO would see that the firm was behind expectations, a purchase of a small firm would often allow reaching the goals.

Crisis followed crisis in the early 2000s and the alternative of M&A became more mainstream. There was even talk about a "buy me" strategy of Start-Up companies that were never looking for the market but rather intended from the ideation stage to be purchased.

According to 2022 statistics of Start-Up Nation Central regarding the Israeli market, one that is quick to react to changes in trends, the numbers are about back to standard. In 2022 there were 89 M&A (first exit) and 17 IPOs and SPACs (normal range between 10 to 20) (Start-Up Nation Central Annual Report, 2022). The actual numbers are of a lesser interest at the moment. Rather more important is the difference in scale which is the main topic to be discussed. That difference in scale is more than just quantitative, it indicates a real difference in strategy. The expected end point may dictate the strategy to be used in order to successfully move to the next phase.

It is easy to claim that this difference stems from avarice, lack of will to work hard – very strange claim for entrepreneurship, or other policy or environmental reasons. Whatever these are, in this book another explanation is offered – that there are inherent reasons for the preference of firms for M&A over IPO or as we would term it, the move towards a complete company.

In order to substantiate this claim, the discussion will start with understanding the difference between a start-up and a complete firm. The nature of the Start-Up activity and mainly that of the Technological Start-Up, will be described and analyzed as well as other types and finally that of the large firm. The fact that there are gaps in the capabilities of the management of each entity and the fact that the move from Strat-Up to Complete firm is not just of growth (size) but rather more complex – of new types of activities and personnel that require different styles of management and a different set of capabilities in the management team.

A preference for M&A over natural evolution towards a complete company has a significant economic effect on a macro level. Relatively small economies with a limited number of large firms that can acquire the new firms and keep them localized, are doomed to "lose" the Start-Ups in favor of multi nationals. That would mean that jobs in manufacturing, marketing and other disciplines which would grow in the complete firm, had it stayed local, would be lost in favor of the multi-national, which we already assumed is not local. These, first tier peripheral jobs, are normally those that assist the relatively weaker sectors in the socio-economic structure of the local economy, and therefore losing them hurts double. Not to mention the second round of peripheral jobs, such as the food industry, office supplies and others that could have benefited from a growing complete company.

One may also look at the role of venture capital, as a market creator for Start-Up firms, which as a market creator would push firm towards M&A, but this is a partial picture at best. In the 1990s when Start-Ups were doing IPOs regularly, the venture capitals were driving that market as well. Current preference for M&A, if it exists as a policy in venture capital, is due to the difficulties mentioned about, and the risks they entail.

In order to make the discussion more effective the book is constructed in the following way.

The book starts with a general literature survey that aims more to present the daunting scale of relevant literature and the main sectors it refers to (Chapter 1).

The discussion of Technology Start-Ups (TSUs) in this book, will focus on the characteristics that differentiate them from other forms of firms and of other activities (Chapters 2-3); how they differ from other "mature" firms (Chapter 4); what is the nature of the gap to be crossed over from a TSU to a mature firm (Chapter 5); followed by a description of the ways to overcome the gap (Chapter 6); and the assistive tools in hand to facilitate that move (Chapter 7). There is also a case study presented, of a less then orthodox firm, making the move in a unique way (Chapter 8). That discussion is followed by a more general reference to the growth of the firm as a learning process (Chapter 9). While each chapter ends with a chapter summary, the book ends with a summary of the topics presented.

Amiram Porath
APC, Israel

REFERENCE

Start-Up Nation Central Annual Report. (2022). https://finder.startupnationcentral.org/snc-2022-report

Introduction

There is a growing discussion regarding the nature of start-ups. The discussion is becoming complex as there are commercial start-ups, companies in most cases, but there are also social oriented start-ups; the companies are also divided into business start-ups, such as service firms, shops etc. and technology-oriented start-ups that are the main topic of discussion in this book. Thus, the discussion entails different legal entities, but also different activities aimed at different types of goals.

The discussion will avoid as much as possible the basics of the nature of the firm, assuming that the interested reader either is already well familiar with the relevant theories or intends to complete the reading elsewhere.

The discussion will be as objective as possible, but as an experienced entrepreneur, I cannot avoid having some preference in the offered solutions. The reader is of course entitled at the same time to disagree, and have a crystal clear, strong, and substantiated, different opinion.

This is not a cookbook, and there can be more than one solution to the problems described below. Therefore, the practical reader, may use which ever seems most appealing, or try a new solution. The move from TSU to a mature firm as we shall see is not an exact science and has as much to do with sociology, culture, and the human spirit as with any technology, and thus is open to interpretations on different aspects.

Readers with interest in developing assistive tools for the move from TSU to mature firm, should remember that while many of the problems discussed below are shared by all TSUs, as they stem from the same source, not all solutions can be alike, and that the solutions have to fit the relevant legal and social system as well as the culture of the environment in which the mature firm is intended to operate.

Having discussed the book up until now, it is time to go ahead and start the discussion, enjoy the reading!

Chapter 1
Literature Review

ABSTRACT

In this chapter the literature related to the topics discussed will be presented in a concise rather than comprehensive form. The reason is to allow the readers to progress to the main topic of discussion without delays. Also, one has to bear in mind that the topic of innovation is multi-disciplinary, and the topic of this book is extremely so. The topics presented in the literature review will present the scope of topics related to the move from start-up to a complete firm. As they encompass the main aspects related to the operation of the firm, they cover a wide area of research and threaten to distract the reader from the topic at hand. Therefore, for each of the topics presented, a skeleton of literature review will be presented mainly to offer a starting point to those interested to pursue a specific topic rather than to present a comprehensive picture of the literature.

DOI: 10.4018/978-1-7998-5685-6.ch001

INTRODUCTION

The purpose of this book is to describe a part in the life of firms that has received less attention than other chapters in the life of firms.

With the advent of the Start-Up firms, the growing awareness of Innovation and what it can do for the economy, researchers in academy, consultants, managers, and government officials at different levels of government develop support tools for innovation.

In other books (Porath, 2010; 2017), I have discussed the supportive tools for the encouragement of Start-Ups. Whether these are private tools such as Angel Investors, and VCs, or public tools, like national and super-national programs their impact, and contribution to the economy, as well as the start-up companies themselves, have been well discussed. My intention is not to declare that the topic of start-ups and support tools is no longer interesting for research, but rather to point to another chapter in the firm's life, that requires more attention than it has received so far. That is the transfer from a start-up to a complete firm. I deliberately do not relate to matureness, or to a specific age (counted in years and months) but rather, to the structure of the firm.

This will be further explained below.

In the literature review below, we shall start with a review regarding Start-Ups (SU), as this seems to be unavoidable. The main concern of Sus is financing their activity so the next section shall relate to the literature regarding SU financing. In order to be able to understand the changes the SU undergoes in the chapter we are discussing in this book; we shall next look at Organizational culture and Organizational culture. The literature review for all these topics shall focus on SUs, not the general aspects, in order to make the literature review as pertinent as possible to the main discussion in the book.

Other topics and sources will be mentioned in the different chapters of the book, whenever appropriate and conducive to the clarification of the discussion at a specific point. Following the maxim of a former teacher of mine: limited number of relevant sources shows the ability to right in a relevant way, a long list of scarcely relevant resources declares an ability to read.

Hopefully, the intention is clear, and it is time for execution.

LITERATURE REVIEW

The main areas of relevant literature are:

The firm and the theory of the firm. This topic is focused on the structure and fit of the firm to the environment as well as the growth and development of the firm. It therefore includes also issues relevant to the organs in the firm and their capabilities.

This leads to the next topic, the one of start-ups. It deals with the different types of start-ups and their role within the research of innovation. Here the research deals with the development of the start-ups as a process of the growth in the capabilities of the different organs of the firm. It is, therefore, close to the topic of learning in start-ups.

The development of the start-ups raises the last major topic which is the support strategies and tools, on local, national and international levels. The tools are derived from the strategies and require their own fit to the specific environment – the relevant ecosystem, which is included in the discussion.

Therefore, the literature review below will give some starting points and will relate to the main topics mentioned here but is intended to invite the reader to further read on the topic that the reader find most interesting.

The Firm and the Theory of the Firm

There is a huge body of publications in that area of research from the publication of Penrose (1959) setting the tone of the research for decades to follow it, and up to the latest publications regarding lean organizations, their structure performance and role. That area of research aims to understand the theory behind the firms (Baum, Locke, and Smith, 2001;), the role of the firms (Folz, 2003;), how they try to construct their structure and internal processes to fit their role, goals and environments(Brown, and Eisenhardt, 1998; Calantone, Cavusgil, and Zhao, 2002; Porter, 2013; Owens, and Fernandez, 2014), the decision-making process that influence the firms (BÉNABOU, 2013; Brunkhorst, 2020), and related topics (Denison, 1990; Jaskyte, 2004; Denison, Janovics, and Young, 2005; Jiménez, and Sanz-Valle, 2011; Mustafa, Lundmark, and Ramos, 2016).

Start-Up Growth and Development

List the sources and their relevance to the topics). As this book aims to give an overview rather than list all the publications up to this point, the following description leaves a lot of the relevant literature unmentioned – just a taste here. The interested reader is invited to search further. The start -up literature mentioned in this book can be divided into the following sub-areas of research:

The Start-Up and the Entrepreneurship

This area of research involves several subtopics. The topic of the **start-up** may include the issues related to the:

- Formation of the start-up (Noy, and Ellis, 2003a,2003b; Yin, and Lou,2018; Santi, and Neves, 2020);
- The function and structure of the start-up – both issues are interrelated (Samila, and Sorenson, 2011; Puri, and Zarutskie, 2012; Sarma, and Marszalek, 2019);
- The growth of the start up all the way to the end of the start-up and the move into the next phase (Barr, and Stimpert, 1992; Nason, and Wiklund, 2018; Porath, 2010).

The topic of **Entrepreneurship** may involve many issues relating to:

- Entrepreneurship (Delmar, and Shane, 2006; Almeida, and Miguel-Oliveira, 2022);
- Entrepreneurs both regarding characteristics and role in the firm and its formation (Entrialgo, and Iglesias, 2017; Kang, 2018);
- The entrepreneurship ecosystem and the interaction with it (Storey, 1991; Hughes, and Morgan, 2007; Hasliza, 2014; Tsegay, Shi, and Wu, 2016);
- Other topics.

These two main topics have increased in research volume since the 1990's and the relevant literature is too vast to cover it sufficiently here even for the sub-topics, or issues mentioned above. The sources presented here are based on personal selection, and the reader is invited to use them both as starting points, and as sourced for keywords for further research (Gompers,

1995; Zahra, 1995; Zahra, Jennings, and Kuratko, 1998; Yiu, and Lau, 2008; Zhang, and Jia, 2010; Taleb, 2012).

Support Strategies

The topic of support strategies relates mainly to the ecosystem and mechanisms created by governments. These may be local, regional, national, bi-national, or multi-national. This topic encompasses also national innovation, R&D, and other support measures, schemes, programs and institutions created in order to further these policies (Lerner, 2000; Audretsch, and Keilbach, 2004; Cohen, 2006; Audretsch, and Belitski, 2016; Porath, 2017; Shiller, 2017).

There has been a growing understanding in economic and policy making circles, that innovation is a growth engine for the economy (Luukkonen, 1988; 2000; Schneider, 1998; EU, 2000;2003; Asheim, Smith, and Oughton, 2011; Feld, 2012; Ács, Autio, and Szerb, 2014; Scalera, 2014; Spigel, and Harrison, 2018; Korbet, and Analyst, 2019), and therefore it should be encouraged by all the entities responsible for economic growth. Once again, this has created a large volume of theoretical research dealing with strategic aspects of such policies, and on the practical research side, a tendency to publish evaluations and comparisons of different programs, institutions, ecosystem analysis etc., aimed at the better understanding of the impact and interrelations of such entities and their goals (Porath, Ramos, and Rahman,2015; Gabai, Ehrlich, Abdel Razek, and Krieger, 2020).

The literature here, is comprised mainly of both academic papers and books, as well as analyses published by official bodies, (ministries, EU etc.) of different aspect related to the topic.

The Gap

Very little specific literature has been found to cover that issue. Most of the literature seems to be concentrated on other "gaps" in the value creating chain of the products, or the formation of the firm. Most of that, is located in the start-up and entrepreneurship section above. In fact, it is that scarcity of literature that brought about the need to write this book.

If Israeli entrepreneurial age (as start-up nation) can be safely assumed to start at 1989-1990, I have had to fortune to participate in the very early stages of it (starting my own first start-up company in 1992), and to remain active in the field through all its changes until now. I have also had the fortune to form TSUs in technology incubators and outside, to work for Technology Transfer companies of two major universities, to participate in an attempt to form a private VC, as well as technology Transfer government-supported company, and to manage for six years three national research support programs. Somewhere in between all that I had the good fortune to work also for large industrial companies. This vast activity including a PhD in management with a thesis regarding the cooperation in R&D between academy and industry, has made me very aware of the gap between a SU and a complete firm.

The scene has changed vastly in the last 30 years. During the 1990's the main model for start-ups was the building of the firm, with the hope of reaching an IPO quickly, before the initial investment run out. The crisis in 2001-2002 put an end to that, in a general way, and the new model replacing it was the M&A. IPO's could no longer be counted upon, as the initial goal of the SU, and entrepreneurs started to build their firms with the intention of leading to an M&A - the "buy me" strategy. The SU no longer had to show that it could be financially viable – be ready to function like a complete company after the IPO, it had now to show that it created enough value in order to raise more funds or be integrated into a larger corporation. It became more important to advance along the value chain, than to create the bases for a complete company.

This also created a shift in the strategy of VCs, aiming for a new exit strategy. The VCs now supported also the creation of value rather that the building of a complete company ready for launch into the world on its own.

The above is a generalization and thus of course may have many exceptions. Like all trends, they take time, some adopted them faster than other, but in a general way, this is a major shift that occurred in the early 2000's. There were of course M&As before that period, and there are still IPOs going on, but this is the general frame of mind.

CONCLUSION

The literature review covers the basic topics mentioned in the book with the aim of arousing the readers interest in further readings according to their topics of interest.

The transition from a start-up to a complete company, or to an alternative route requires further research. It is related to research topics such as innovation, the innovation ecosystem, micro and macroeconomics and the theory of the firm. As it deals with start-ups and especially with technology-based start-ups and leads to the full company it can actually touch most of the business and management research areas. However, in this book and in the literature review I have chosen to focus on the main aspects of relevant theory.

As mentioned above, the relevant fields themselves cover a vast volume of publications which could not be covered in a single chapter, and therefore the readers are invited to delve into the vast volume of knowledge and select their own reading. The transition itself, requires not only further research and case study development, but would also benefit from an extension of the theory of the firm to cover this special transition. A better understanding of the transition would assist future decision makers in developing strategies and plans to further the development of their economy.

And now, it is time to move to the specific discussion in this book.

REFERENCES

Ács, Z. J., Autio, E., & Szerb, L. (2014). National Systems of Entrepreneurship: Measurement Issues and Policy Implications. *Research Policy*, *43*(3), 476–494. doi:10.1016/j.respol.2013.08.016

Almeida, F., & Miguel-Oliveira, J. (2022). The Role of Intrapreneurship in Portuguese Startups. *Periodica Polytechnica Social and Management Sciences*, *30*(1), 70–79. doi:10.3311/PPso.17796

Asheim, B. T., Smith, H. L., & Oughton, C. (2011). Regional Innovation Systems: Theory, Empirics and Policy. *Regional Studies*, *45*(7), 875–891. doi:10.1080/00343404.2011.596701

Audretsch, D., & Keilbach, M. (2004). Entrepreneurship Capital and Economic Performance. *Regional Studies*, *38*(8), 949–959. doi:10.1080/0034340042000280956

Audretsch, D. B., & Belitski, M. (2016). Entrepreneurial Ecosystems in Cities: Establishing the Framework Conditions. *The Journal of Technology Transfer*, *42*(5), 1030–1051. doi:10.100710961-016-9473-8

Barr, P. S., Stimpert, J. L., & Huff, A. S. (1992). Cognitive change, strategic action and organizational renewal. *Strategic Management Journal*, *13*(Special Issue), 15–36. doi:10.1002mj.4250131004

Baum, J. R., Locke, E. A., & Smith, K. G. (2001). A Multidimensional Model of Venture Growth. *Academy of Management Journal, 44*(2), 292–303. doi:10.2307/3069456

Brown, S. I., & Eisenhardt, K. M. (1998). *Competing on the Edge*. Harvard Business School Press.

Brunkhorst, H. (2020). "Everything is now in order" – Groupthink, ideology and practical-critical activity in the European financial crisis and beyond. *Journal of European Public Policy, 27*(9), 1406–1423. doi:10.1080/13501763.2020.1773515

Calantone, R. J., Cavusgil, S. T., & Zhao, Y. (2002). Learning Orientation, Firm Innovation Capability, and Firm Performance. *Industrial Marketing Management, 31*(6), 515–524. doi:10.1016/S0019-8501(01)00203-6

Cohen, B. (2006). Sustainable Valley Entrepreneurial Ecosystems. *Business Strategy and the Environment, 15*(1), 1–14. doi:10.1002/bse.428

Delmar, F., & Shane, S. (2006). Does Experience Matter? The Effect of Founding Team Experience on the Survival and Sales of Newly Founded Ventures. *Strategic Organization, 4*(3), 215–247. doi:10.1177/1476127006066596

Denison, D. R. (1990). Corporate Culture and Organizational Effectiveness. New York: John Wiley & Sons.

Denison, D. R., Janovics, J., & Young, J. (2005). *Diagnosing Organizational Cultures: Validating a Model and Method-Working Paper*. Denison Consulting.

Entrialgo, M., & Iglesias, V. (2017). Are the Intentions to Entrepreneurship of Men and Women Shaped Differently? The Impact of Entrepreneurial Role-Model Exposure and Entrepreneurship Education. *Entrepreneurship Research Journal, 20170013*(1), 1–14. doi:10.1515/erj-2017-0013

EU. (2000). *Lisbon European Council 23 and 24 March 2000*. https://www.europarl.europa.eu/summits/lis1_en.htm#1

EU. (2003). *Communication from the Commission; Investing in research: An action plan for Europe*. COM (2003) 226 final. Available online at: https://ec.europa.eu/invest-in-research/pdf/226/en.pdf

Feld, B. (2012). *Startup Communities: Building an Entrepreneurial Ecosystem in Your City*. Wiley. doi:10.1002/9781119204459

Folz, J. M. (2003). Managing two brands for success: Peugeot and Citroën. In B. Gottschalk & R. G. Kalmbach (Eds.), *Markenmanagement in der Automobilindustrie*. Gabler Verlag. doi:10.1007/978-3-322-92081-2_16

Gabai, U., Ehrlich, E., Abdel Razek, K., & Krieger, N. (2020). *High-Tech Human Capital Report 2019*. Israel Innovation Authority and Start-Up Nation Cantral.

Gompers, P. A. (1995). Optimal Investment, Monitoring, and the Staging of Venture Capital. *The Journal of Finance*, *50*(5), 1461–1489. doi:10.1111/j.1540-6261.1995.tb05185.x

Groupthink, B. (2013). Collective Delusions in Organizations and Markets. *The Review of Economic Studies*, *80*(2(283)), 429–462.

Hasliza, A. (2014). The Growth of Innovative Performance among SMEs: Leveraging on Organisational Culture and Innovative Human Capital. *Journal of Small Business and Entrepreneurship Development*, *2*(1), 107–125.

Hughes, M., & Morgan, R. E. (2007). Deconstructing the Relationship between Entrepreneurial Orientation and Business Performance at the Embryonic Stage of Firm Growth. *Industrial Marketing Management*, *36*(5), 651–661. doi:10.1016/j.indmarman.2006.04.003

Jaskyte, K. (2004). Transformational Leadership, Organizational Culture, and Innovativeness in Nonprofit Organizations. *Nonprofit Management & Leadership*, *15*(2), 153–168. doi:10.1002/nml.59

Jiménez, D. J., & Sanz-Valle, R. (2011). Innovation, Organizational Learning, and Performance. *Journal of Business Research*, *64*(4), 408–417. doi:10.1016/j.jbusres.2010.09.010

Kang, H. D. (2018). A Start-Up's R&D Stages and the Evolution of Financing Sources: Evidence from the Biotechnology Industry. *Entrepreneurship Research Journal*, *20170159*(3), 1–19. doi:10.1515/erj-2017-0159

Korbet, R., & Analyst, M. (2019). *The State of the Israeli Ecosystem in 2018*. Start-Up Nation Central.

Lerner, J. (2000). The government as venture capitalist: The long-run impact of the SBIR program. *Journal of Private Equity*, *3*(2), 55–78. doi:10.3905/jpe.2000.319960

Luukkonen, T. (1988). The difficulties in assessing the impact of EU Framework Programs. *Research Policy*, *27*(6), 599–610. doi:10.1016/S0048-7333(98)00058-4

Luukkonen, T. (2000). Additionality of EU Framework Programs. *Research Policy*, *29*(6), 711–724. doi:10.1016/S0048-7333(99)00041-4

Mustafa, M., Lundmark, E., & Ramos, H. M. (2016). Untangling the Relationship between Human Resource Management and Corporate Entrepreneurship: The Mediating Effect of Middle Managers' Knowledge Sharing. *Entrepreneurship Research Journal*, *6*(3), 273–295. doi:10.1515/erj-2015-0004

Nason, R. S., & Wiklund, J. (2018). An Assessment of Resource-Based Theorizing on Firm Growth and Suggestions for the Future. *Journal of Management*, *44*(1), 32–60. doi:10.1177/0149206315610635

Noy, E., & Ellis, S. (2003a). Risk: A neglected component of strategy formulation. *Journal of Managerial Psychology*, *18*(7), 691–707. doi:10.1108/02683940310502395

Noy, E., & Ellis, S. (2003b). Corporate Risk Strategy: Does it Vary Across Business Activities? *European Management Journal*, *21*(1), 119–128. doi:10.1016/S0263-2373(02)00159-7

Owens, T., & Fernandez, O. (2014). *The Lean Enterprise: How Corporations Can Innovate Like Start-Ups*. Wiley. doi:10.1002/9781118854006

Penrose, E. T. (1959). *The Theory of the Growth of the Firm*. John Wiley.

Porath, A. (2010). *Collaborative Research in Management: Inside out*. SAGE Publications.

Porath, A. (2017). *Supporting Innovation Through National R&D Programs: Emerging Research and Opportunities*. IGI Global.

Porath, A., Ramos, I., & Rahman, H. (2015). Different approaches to technology transfer by government and academy. *World Review of Entrepreneurship, Management and Sustainable Development*, *11*(2/3), 296–307. doi:10.1504/WREMSD.2015.068585

Porter, M. (2013). *The case for letting business solve social problems*. https://www.ted.com/talks/michael_porter_the_case_for_letting_business_solve_social_problems

Puri, M., & Zarutskie, R. (2012). On the Life Cycle Dynamics of Venture-Capital and Nonventure-Capital Backed Companies. *The Journal of Finance*, *67*(6), 2247–2293. doi:10.1111/j.1540-6261.2012.01786.x

Samila, S., & Sorenson, O. (2011). Venture capital, entrepreneurship, and economic growth. *The Review of Economics and Statistics*, *93*(1), 338–349. doi:10.1162/REST_a_00066

Santi, C., & Neves, G. T. (2020). Startups and prominent business. In A. G. Rodrigues (Ed.), *New and Future Developments in Microbial Biotechnology and Bioengineering* (pp. 209–223). Elsevier. doi:10.1016/B978-0-444-64301-8.00010-X

Sarma, S., & Marszalek, J. (2019). New Venture Growth: Role of Ecosystem Elements and Prior Experience. *Entrepreneurship Research Journal, 10*(1). . doi:10.1515/erj-2018-0215

Scalera, F. A. (2014). Internationalisation Strategies in the Automotive Sector. The PSA Peugeot Citroen and Fiat Auto Groups: A Comparison between the two Models. *JIEB*, (2), 28–37.

Schneider, S. C. (1998). National Vs. Corporate Culture: Implications for Human Resource Management. *Human Resource Management*, *27*(2), 231–246. doi:10.1002/hrm.3930270207

Shiller, R. J. (2017). Narrative Economics. *The American Economic Review*, *107*(4), 967–1004. doi:10.1257/aer.107.4.967

Spigel, B., & Harrison, R. (2018). Toward a Process Theory of Entrepreneurial Ecosystems. *Strategic Entrepreneurship Journal*, *12*(1), 151–168. doi:10.1002ej.1268

Storey, D. J. (1991). The Birth of New Firms – Does Unemployment Matter? A Review of the Evidence. *Small Business Economics*, *3*(3), 167–178. doi:10.1007/BF00400022

Taleb, N. N. (2012). *Antifragile. Things that gain from disorder*. Random House Trade Paperbacks.

Tsegay, B. K., Shi, C. S., & Wu, W. W. (2016). The Impact of Innovation Strategy on Organizational Learning and Innovation Performance: Do Firm Size and Ownership Type Make a Difference? *South African Journal of Industrial Engineering*, *27*(1), 125–136.

Yin, B., & Lou, J. (2018). How Do Accelerators Select Startups? Shifting Decision Criteria Across Stages. *IEEE Transactions on Engineering Management*, *65*(4), 574–589. doi:10.1109/TEM.2018.2791501

Yiu, D., & Lau, C. (2008). Corporate Entrepreneurship as Resource Capital Configuration in Emerging Market Firms. *Entrepreneurship Theory and Practice*, *32*(1), 37–57. doi:10.1111/j.1540-6520.2007.00215.x

Zahra, S., Jennings, S., & Kuratko, D. (1999). The Antecedents and Consequences of Firm-Level Entrepreneurship: The State of the Field. *Entrepreneurship Theory and Practice*, *24*(2), 45–66. doi:10.1177/104225879902400205

Zahra, S. A. (1995). Corporate Entrepreneurship and Financial Performance: The Case of Management Leveraged Buyouts. *Journal of Business Venturing*, *10*(3), 225–247. doi:10.1016/0883-9026(94)00024-O

Zhang, Z., & Jia, M. (2010). Using Social Exchange Theory to Predict the Effects of High-Performance Human Resource Practices on Corporate Entrepreneurship: Evidence from China. *Human Resource Management*, *49*(4), 743–765. doi:10.1002/hrm.20378

Chapter 2
The R&D-Based Start-Up

ABSTRACT

In this chapter the concept of the difference between a technological start-up and a complete firm is introduced. The discussion explains the characteristics of the technological start-up (TSU) and the reasons these have developed as part of a strategic fit of the TSU to its specific tasks and environment. The discussion relates that the characteristics of the TSU, structure, culture, and activities have an impact on management capabilities and their development. The discussion along these three aspects also denotes the literature in the field in support of the claims made. However, the discussion is not entirely theoretical but includes also references to practical aspects.

DOI: 10.4018/978-1-7998-5685-6.ch002

INTRODUCTION

Start-Ups (SU) and Technological Start-Ups (TSU) have been of interest for some time for policy makers, as they have been recognized as economic growth engines (EU, 2003).

The interest in the TSUs and the environment supporting their formation and growth have therefore also been the topic of wide research as well as of practical interest for some time (EU, 2003; Porath, 2010). However, while in former years the main obstacles for research regarded the formation of SUs and the supporting environment, with the maturing of the sector of TSUs a new issue is slowly becoming interesting – the move or transformation from a TSU to a complete firm. In order to analyze the transformation, it is first important to understand the special characteristics of the TSU, and how the TSU differs from the classical idea of a complete firm (Penrose,1959). In this chapter, the TSU characteristics, their origin and how they relate to the environment are discussed. Thus, the foundation for the main discussion of the book is presented in this chapter.

Challenges the Start-Up Faces

The TSU is a special entity in the firm group. While a mature, or fully developed firm, a complete firm, would contain the funding generation activities and units (marketing and sales), the management part (administrative, financial etc.), and the production (including services) as well as support activities sections, the TSU normally would lack some of the sections, or have them at a less developed stage than a mature firm.

There are many different ways to define a start-up, but for our discussion here, we shall define the TSU as follows: A Technology based company, focused on the development of its technology and products, supported in the main part by investment (equity or non-equity based) and not sales[1].

The TSU therefore is a company, that would mean for our purposes here, a for profit organization, which however, at the current stage we are dealing with, is not showing any profit, or better defined as one that is not generating enough income to be fully supported by the sales of its product.

Imagine the classic view of such a TSU. Two guys, long haired and bearded working on a new electro-optic device in the garage of the home of one of them. They manage to convince an early-stage investor (Angel) to invest some funds and recruit another team member to write the software they need. This company, even when it grows to 10 members, will not have all the sections of a mature firm. It will rarely sell anything, and will recruit mainly development team members to progress in the R&D. It may recruit some administrative stuff, especially as it grows but will not recruit a full team for marketing and sales, finance, or post sale support.

While a complete mature firm will need to be able to stand on its own two legs (or how many it may require) in order to survive, the TSU will need to show only progress in the R&D in order to generate the next phase of investment. At a later stage, it may need to show the ability to go into the market, but until that phase is reached, it is mainly the R&D department of the future firm it hopes to grow into. There is one more aspect and that is the cultural-organizational aspect. The hierarchy in a TSU is flat, the relations informal, and the job descriptions mainly to do whatever is needed. That all do what is needed, is especially true for the founders-entrepreneurs, who are hand –on regarding every aspect of the new firm.

Up to this point we have identified the following important special identifiers of the TSU:

- It is the future firm's R&D activity with some add-ons.
- It is not based on or fully oriented towards generating sales – the required resources are raised as investment (equity based or not), based on R&D progress.
- The link to the market, is based mostly on assumptions and market surveys, not on real sales.
- Special cultural organizational characteristics.

There are therefore some challenges which stem from the above-mentioned identifiers:

- How to manage the operation of the TSU when it is not s fully sustainable operation – it cannot create its own required resources?
- How to make sure that the firm survives and even grows into the mature firm without the required resources?
- What should the focus of the firm be? More R&D or develop some of the support activities?
- How to correctly grow the firm and develop the different sections / activities so as to make sure that it develops into the mature firm of the future?

- How to make the organizational cultural shift and create orderly job description, a rigid organizational structure, and manage all this remotely through delegated authority, while keeping the founders involved and happy? When to do this? This is the main discussion point of this book.

These are not trivial challenges, nor are they simple to solve. However, they are the challenges that the managers of TSUs face on a daily basis and need to solve regularly in order to make sure their firm survives.

Whenever a new firm reaches the news, due to a merger with a giant, an IPO that proclaims it a huge success, or similar, we feel that the TSU world is full of wonder and success. It is important to remember that most TSUs do not reach a happy end. It is important to recall, that the challenges mentioned above, often are too much for the TSU and it succumbs to the pressure and collapses. Or just fails to make the next round of investment and fades.

However, the TSU has similarities with the mature firm. The main one is of course the generation of resources. Both types cannot survive without the generation of resources. While the mature firm makes the strongest effort to generate the required resources for its operation, under which ever strategy it chooses, the TSU also needs to make sure it creates the best opportunities for the generation of resources – the next investment.

Another similarity lays in the need to progress. While the mature firm will measure the progress in terms of increased sales, increased revenues, market share, share value etc., the TSU will measure the progress in shortening the time to market, in advancing along the development stages (e.g., moving from one TRL[2] to another). Other measurements, such as number of people employed, or funds used, do not necessarily measure progress, just expenditure.

In that regard, one of the major characteristics, to be also discussed below in the culture section, is the personnel in the TSU. Employment in the TSU is mostly limited to technical stuff, will very little support personnel. These people will be technologically trained, and of their cadre will also be the ones who are at least partly motivated by the challenge, as employment in the TSU is rarely long-term security. People seeking long term employment security are inclined to work in mature firms. The nature of the TSU, where the future is in-secure, tends to attract people seeking the feeling of achievement, and the excitement of creating something new. That type of personnel will also tend to be innovative, thinking outside the box, and seeking solutions over sticking to known answers. They will also require leadership over regulation in order to progress.

Organizational Culture Characteristics

One of the characteristics of a TSU is the culture. There has been a lot of discussion regarding the TSU culture, referring to the long hours, the lack of formality and the atmosphere of creativity and out of the box thinking (Tsegay, et all, 2016; Calantone, et all, 2002).

It would appear that at least for some, this is the main appeal of the TSU, that is to say the less than formal atmosphere, that would seem to fit the friendly environment that the Y generation likes.

The chaotic culture, lack of formality is conducive to creativity, which is the basis of the innovation in the TSU (Denison, 1990; Denison et all, 2005). The lack of formality allows the raising of new ideas and concepts, and at the same time creates the required incentive for the strenuous life in the TSU.

The corridor discussion, the "brain storming" meetings and the ability of each of the workers to speak up are part of the innovation process in the TSU. The important aspect of the TSU is the sacrifice of the efficiency for the innovation / creativity in the TSU (Hasliza, 2014).

The corridor discussion which could seem to be an unimportant activity with low contribution to the creativity and the innovation of the firm is in fact far from it. The important aspect of the creativity in any organization in the current world is the ability to mix disciplines not only during the solution stage of activity, but just as important (at least) at the question definition stage. The corridor discussion mentioned here, is a representative of any unscheduled meeting in which problems are mentioned and possible solutions are discussed. But even at the phase of explaining the solution, to someone from a different department with different outlook at things (human resources versus engineering is always interesting) would normally result in a better understanding of the different aspects of the problem than before, for the party explaining the problem. There is much discussion regarding the creation of effective teams that would avoid group thinking (BÉNABOU, 2013; Shiller, 2017; Brunkhorst, 2020).

The corridor discussion is a de-facto multidisciplinary team formation, unplanned and unscheduled, but nonetheless very effective.

When explaining the problems to someone from a different discipline the problem has to be explained from the fundamental aspects and also explaining the basic assumptions – this part is normally skipped when discussing the problems with people from the same discipline as they tend to have the same basic assumptions – creating a group thinking issue. The explanation of the basic assumption the new clear fundamental definition of the problem is often a de-clattering stage for the explaining person and results in a new and clearer understanding of the problem for him. This leads many times to a new definition of the problem, simpler, more concise, or otherwise clearer and this in many cases in the conducive step towards a solution.

In fact, for companies that are small in size, where most people know each other so that the discussion is open and easy, this corridor discussion can be more effective than the long team meetings in which everyone is busy with stating their opinions. Here the nature of the meeting, haphazard and short time, requires some reduction in the definition of the problem, focusing on the main important aspects. The questions asked are to the point, to make things clearer for the listener, but also in this way to the explaining party, and the short time and lack of additional participants are in fact assisting in getting to the final point.

A culture of openness and lack of formality in which the CTO and the newest member in sales can talk freely calling each other by their first name, and in fact knowing each other, is a clear benefit to the TSU, regarding the effectiveness and in fact the happening of the corridor discussion.

Another interesting tool is the "brainstorming" meetings. This is a tool that was very popular in the late 90's and became less popular when it became clearer that you needed to understand group management in order to get an effective "brainstorming" meeting (Mustafa, et all., 2016). However, when solutions "out of the box" are required that could be a very effective tool. One should bear in mind, that it still requires the knowledgeable management, and the time resource to be allocated for it. I find that this tool is more suited for complicated issues, while the corridor discussion is better fitted for simper issues – the time element dictating the fit, but that is not a rule, and both can be used for all types of problems. The main difference is normally in the cost. The most important resource for a TSU is time and the corridor discussion, engaging normally two or three participants, being short by nature is less consuming of the time resource. The "Brainstorming" meeting would normally consume the time of more than three people and would require more time to allow them to raise ideas and issues and have them discussed (Almeida and Miguel-Oliveira,2022).

Regarding time as the most precious resource for a TSU, the best way to demonstrate it is via accelerators. As the duration a TSU can spend in an accelerator is limited, that resource, time, cannot be replaced (Yin and Lou, 2018).

As already mentioned, both are important and should be employed. By the way, to benefit from corridor discussions people from different disciplines should be going throw the corridor – easier to do with a small firm where all sit close to each other, or share the same coffee corner, than in larger organizations, where sometimes the whole building is just one department.

As mentioned above the personnel employed in the TSU will tend to be less prone to permanent employment and more achievement motivated. That is not to say that the employees are all risk lovers, nor even that all the entrepreneurs are such risk seekers (Noy and Ellis, 2003a; 2003b), but rather that they prefer the sense of achievement derived from overcoming problems other could not, doing it in a new way, or creating something new, to the sense of security, of a guaranteed employment in a large mature firm.

If anything can be said to be common to all TSUs it is the sense of insecurity on the one hand with the excitement and interest on the other (Barr, et all, 1992), there are never dull moments.

One of the important characteristics in a TSU, especially at the beginning and with a diminishing impact as maturity sets in, is the way personnel is employed. Whether the TSU is the recruiting agent (in many cases based on a friend bringing a friend), or an external agency is doing it for the TSU, partly due to its, culture of chaos and partly due to its fast growth rate, two characteristics are so strong, they can be used as indicators in defining the TSU:

1. Employees are given responsibilities and job descriptions are defined according to the capabilities of the employees;
2. The first recruited employee for each new activity will be a veteran of such an activity from the outside and will bring the work methodology with him.

These requires some explanation. The first point means that in cases where positions are required either to adhere to some standard, (e.g., ISO), or a new demand arises from the activity (e.g., purchasing manager), the first tendency would be to look to an already existing team member with some relevant experience and task them with the responsibility. This would lead to the creation of strange combination for example: one member may be the GUI development team leader and also the standard adherence coordinator, or the purchase manager for the TSU. In mature firms where there are relevant approval processes and there is normally time (or no choice but to allocate the time), the firm will recruit people for well-defined position once the position had been deemed necessary and has been define. Thus, in TSU there is a tendency to define position based on existing personnel capabilities, while in mature firms, the tendency is to recruit people per their fit to the requirements of well-defined positions.

The second point is more crucial sometimes and can be a barrier to the creation of a local eco-system. For TSUs in order to create their new body, the first people to be recruited are the leaders, in entrepreneurship as well as in technology. That would mean that for each development team the first people to be recruited are those with the experience of leading such teams and in leading similar development projects. These would normally be people with experience and in most environment the scarcest human resource.

It is important to understand that a TSU is not a small copy of a regular, or complete company. The TSU is in fact the R&D department of a regular company with some capabilities added to it for functionality reasons. That is the basis for the unique character of the TSU as well as the difficulties discussed below when making the transit into a mature company.

Management

If the reader is under the impression that the partial nature of the TSU means that it is easier to manage it than a mature firm, or even an R&D department in a mature firm, this is the wrong impression.

The partial nature of the firm removes some of the distractions that management in mature firm would entail, however, at the same time it also removes most of the support the managers in such firms would receive and give each other. The manager of the R&D department does not need to manage the financials of the entire firm, at most the budget of the unit, but certainly not the bank accounts, the relations with the investors and other daily tasks the financial manager takes care off in a mature firm. In the TSU the general manager is tasked with dealing with these responsibilities or at least overseeing them.

While in a mature firm the R&D manager (CTO, Chief Scientist or any other title) may advise the IP manager (internal or external), in the TSU the manager is tasked with dealing with IP issues hands on, as this is in most cases the main asset of the firm and the basis for future investments. It is an essential resource for fund raising and as such demands the direct involvement of the entrepreneurs.

Human resources management, salaries etc., are additional managerial tasks that normally the R&D managers are exempt from dealing with in mature firms. They are involved in the development of the capabilities of their personnel and hiring new employees of course. The daily management and administration are normally dealt with other departments.

It is important to understand that while at least partially some of the above may be out-sourced to external contractors, it is still the responsibility of the management of the TSU to oversee the activities out-sourced, and initially to identify and contract the external parties.

Therefore, the management responsibilities of the TSU manager are no less than those of a manager in a mature firm, however the support management capabilities from other professional managers are lacking. It is important to recall that in most cases, at least initially the TSU will be managed by the entrepreneur, or the entrepreneurial team and will therefore tend to be stronger on the technological capabilities, than on the managerial capabilities. On the other hand, the founders tend to be personally involved in all aspects of their firm. This is hands-on management. There are no advanced communication and reporting mechanisms in place as they are required. This means also that the remote management capabilities are limited in the organization.

While this may seem trivial, there some immediate implications resulting from the management aspects on the development of the TSU, and on the organizational culture, as well as on the direct view of the management of the firm.

Let us consider the simple, and therefore the most difficult scenario. A single, electrical engineer by training, entrepreneur, is forming a new TSU. The entrepreneur raises initial funding for the idea, forms a company and employs two additional workers. The entrepreneur has to oversee the finance and deal with the bank, hire an accountant to manage the accounting and prepare reports, a firm to take care of preparing salaries, and transferring tax payments, social security payments, insurances etc., while directing the R&D activity, dealing with patent attorneys, reporting to the board on a periodical basis, and making sure that no opportunity for further fund raising is missed. He[3] personally knows everything that goes on and is personally involved in all actions.

We clearly understand that the entrepreneur would need not only be a good engineer, a leader inspiring the other workers, but also be able to negotiate, understand and manage (make decisions, select options) in different fields of actions, that are not the typical engineering activity. This is a heavy burden, explaining the long hours, the fact that entrepreneurs of TSUs tend to live on nervous energy and lead in general un-healthy lives.

It is understandable that in order to cope with the decisions demanded from managing such an activity the manager would need to have a wide understanding of different aspects of management, a jack of all trades to speak off, and do so while continuing to lead the R&D activity. There are several ways to manage that difficult task. He would also need to be able to learn new subjects quickly and adapt to new demands.

There is a clear advantage to experience in such endeavors. The manager of a TSU, successful or not[4], will have a lot of experience in how to cope with the different difficulties, will know how to contract and negotiate with the different external contractors, and will therefore be able to avoid a lot of the mistakes that may kill the new TSU. It is for this reason that there is such a strong market for experience entrepreneurs, adding to that fact, that entrepreneurship is addictive like many other adrenaline intensive activities – we get the serial entrepreneurs. So, experience is a partial solution, but it does not solve the time allocations problem – the ability to find enough time to take care of all of the activities sometimes at the same time slot.

One of the common solutions is to have a team of entrepreneurs who know each other and how to work together, and who can share the responsibility burden, based on their past experience. This, however, comes at a price, in the sharing of the ownership of the firm and therefore the resulting financial gains and benefits. The allocating of the tasks stemming from such a solution is that the organizational structure and role definitions are made to fit the capabilities of the management team, and this trait stays on for the future TSU organizational structure. Once a specific setting become a used to solutions, a tradition is formed, making future changes more difficult and emotionally costly.

The management of the TSU is connection intensive. The manager needs to be able to access different people, such as potential customers for marketing information important for the R&D activity, investors of different kinds (e.g., angels, VC's etc.), standardization experts and many others. Acquiring and maintain such connections is a difficult task. Once again, experience would normally supply most of the required connections but probably never all of them. One solution to both the disciplinary issues and the connection dilemma is the Advisory Board (AD).

While the experienced entrepreneur would keep the TSU board members to a possible minimum, both in order to be able to manage it without much help, and to allow it to grow with future investment, the entrepreneur could use the AD as a supplement.

The AD is an organ of the firm that can be used to advise the manager on different issues where decisions or the ability to reach new contacts are required. The advisory board allows recruiting experts without the need to either give them the coveted board seat, reserved to equity investors.

The attraction of being AD members to people that have normally no significant vested interest in the firm, is first of all the personal ties to the entrepreneur but also the membership of an exclusive club, and the informal ties formed by this membership. So, while this organ is laxer in the admission to its membership, it is important to maintain the quality of the members, and make sure that each new member added, brings some additional value, in expertise or connections, but also making sure the new member increase the AD reputation, making it even more attractive than before.

The AD may also serve as a fund-raising tool, by enhancing the prestige of the TSU based on the prestige of the AD members. If such persons are willing to tie themselves to the TSU, and they are serious people, there must be a real value in the TSU. This will explain the publication on the TSU website of the AD membership.

Another aspect many managers adopt in order to cope with the burden of managing the TSU with little help, and contending with the different disciplines required, is the adoption of tried solutions, whenever they see the activity as a side or less critical activity. The managers would tend to favor external service providers they heard from other entrepreneurs' references for or adopt processes and sometimes forms for activities that work in another place. In such cases, one may find that a certain accounting firm is working with several TSUs in which the managers have a common background (coming from the same university, joint military service etc.), that the purchase forms in TSU sitting close together are in fact the same forms and that the purchasing is normally done from the same providers.

In such cases the managers are often ready to relinquish some of the efficiency of managing the resources in order to save time and mind share. The purchasing of office equipment is not critical to the firm, certainly less critical than dedicating the time and mind share to the progress of the R&D process, and therefore even if costlier than an alternative, would be preferable to having the manager optimize the purchase of office equipment. This would need to change once the company has mature enough to have a purchase manager, who would be required to optimize the purchase process.

In fact, for a young TSU all the non-R&D activities are considered superficial to the survival of the TSU and can therefore either be relegated to external service providers, or external tried if not optimal solutions may be adopted as a conscious strategic decision.

The Resulting fit to the R&D Phase

In this section the specific fit of the TSU to the R&D phase is described regarding both the structure (lack of sectors/ divisions in the TSU compared to mature firm) and the culture on its different aspects to the specific R&D intensive stage. The resulting type of entity formed and the possibility of difficulties in moving from this position.

The mature firm, competing in the marketplace with other firms is aiming at optimizing its performance. Maximum efficiency in the production sales and after sales is a must for firms with mature products aiming at maximizing their profit.

Without getting too deep into economics, the profit firms derived from their market activities is the difference between their expanses and their income. The income is derived from the market, which dictates in many cases the possible price for the products / services of the firm[5]. If the price is fixed, in order to maximize profits, the firm have to focus on reducing costs and maximizing the sales potential.

Reducing the costs, includes improved efficiency and this the firms are trying to become more and more efficient, and the same for increasing the sales. The increase efficiency pervades all the aspects of the firm, from the procedures and the culture to the actual way of thinking and the set-of-mind of the managers.

It would therefore not be surprising that the entire firm structure would be aimed at increased efficiency, aiming to make the logistics, production, sales and post sales services, as efficient as possible. The different departments with their internal efficiency, the interfaces among them as well as the external entities involved in the market.

The intended efficiency is not a trifle issue. The organizations that do not manage to achieve enough efficiency are liable to vanish due to depleted resources, as their profit and therefore the resulting financial viability are dependent upon a minimum of efficiency that would allow them the profit.

The other aspect is the completeness of the performance capabilities in the mature firm. As the mature firm is an independent entity, it must be able to produce the resources it requires in order to survive on a long-term basis. Therefore, it requires the ability to produce, market, sell, collect, provide services etc., including the abilities to recruit the resources such as HR, raw materials, and other production inputs, as well as maintain the overhead and indirect inputs for the continuous and regular operation of the firm.

As the firm is required to independently supply all of the above and to do it in an efficient manner so as to keep going, we can refer to it as efficiency and structure for a sustainable long-term existence, it is important that the firm is self-sufficient in that regard.

These two parameters, the efficiency of operations and the need to be self-sufficient for long term sustainability, dictate the culture and the structural characteristics of the mature firm. But what about the TSU?

The Technical Start-Up is a different type of entity. It is in many cases, at least by design, a temporary entity, standing on the brink of decision. The decision to die or to evolve into a mature company. About the nature of that decision more shall be discussed at a later part of this book. The need of the TSU in order to evolve and not die, is to utilize the resources it has to progress in the technological development to the next stage, which will allow it to raise the next phase of the resources, so as to make sure it reaches the development stage in which it is ready to evolve into a mature firm. In order to do that the TSU needs to focus on its technological development, and to make sure that it is focused on that and not distracted to waste its limited resources and miss the ability to raise the next phase of the resources for continuous development.

The TSU therefore is focused on its R&D efforts and limits the other operations to the minimum required for continuous operation. It will limit for example HR activity to the recruitment and payment for the HR it employs, and only deal with HR development and benefits if it obliged to do so by the competition, sectorial tradition, or personal experience of the entrepreneurs. The TSU to all purposes is operating as an R&D department of a mature company, with the crucial capabilities of the firm required for sustainability until the next phase.

Efficiency of the entire operation is not a crucial requirement for the TSU, as the non-R&D activities it performs are normally limited in scope and size compared to the R&D activity. In the TSU, the paramount need is the innovation, the ability to go beyond the state-of-the-art, and establish some long-term strategic advantage[6] in the market. That need and the resulting adapted structure and culture explain the special characteristics of the TSU, compared to those of the mature firm. It is a slim focused and R&D adapted organization, aimed at reaching the next stage in development.

The TSU is mentioned here as a single typical entity, but the TSU is in fact, in many cases, a work in progress, an evolving structure which at least partly matures as it moves from one capital raising round to the next (see Chapter 8 – SeeTRee Ltd. - a case study).

As can be expected, if the TSU is an organ, a part of an organization, it will need to get the support which for the mature firm the other organs, the other departments and unit's supply. In the next section we shall discuss that support and how it can be provided. It is of course, due to the nature of the TSU, provided externally, by the environment[7] in which the TSU operates.

A Healthy TSU Ecosystem

It is important to recall that the TSU is not an island entirely by itself. Presented is the importance of the ecosystem supporting the TSU. The need of the TSU to focus and thus to copy the experience from others, to get access to knowledge, the strength of numbers in getting access to knowledge, finance, the new aspects of co-competitors in TSU environment and the needs of such an environment.

The attractiveness of comparing the TSU to a biology parallel is very compelling. The TSU is in fact, not a complete living entity, but rather an organ, that requires the rest of the organs in order to nourish and sustain it. In the complete firm, all the other organs are in place and therefore sustaining the firm is less of a problem – even for firms that are competing in very competitive markets. The simple test is the source of the resources. In the complete firm the resources are generated through the activity of the firm. The income the firm generates from its activities fuel its further operation. In the TSU it is the investments that fuel the firm. The firm without its investments, as many failed entrepreneurs have found out, dies. The TSU is not capable of creating its own resources and requires "feeding" from external sources.

The impact of this difference is paramount. The TSU therefore lives under the threat of resources running out (Puri and Zarutskie, 2012; Kang, 2018). It is further compelled to make sure that its regular operations, the usual way of doing business, propels it towards the next investment. It is always required to show not only growth, as would also a complete firm be required, but advancement, specifically in the technology. As the key for the next investment would be the technological advancement in the R&D, and to a lesser extent the business advancement – especially in market understanding, and regulation where relevant, an investment in any other activity, on the side of the TSU, is a waste of resources. It is un-effective, as such a side activity would not improve the chances for the next investment, the size of the next investment or the valuation it will be done under.

In addition to dedicating the efforts towards technological advancement, as mentioned above some supportive additional activities are required at this time. These activities support the development process and assure that it is valid as well as effective, and thus conducive to the next invention. Therefore, one of the most critical problems for the TSU is obtaining information and developing the system of information resources (Korbet and Analyst, 2019). The development of the information system, regarding both different aspects of the regulations (legal, fiscal, technical, and other), as well as regarding opportunities and dangers is complex, time consuming and requires a lot of energy and additional resources (Porath, Ramos and Rahman, 2015). The ability to "suckle" from others in this aspect, can impact in a critical way the TSU making the difference between success and failure (Gabai et all, 2020).

To sum the needs differentiating the TSU from the complete firm, relate to the need to be able to make the next investment at all, and preferably under the best terms. Thus, the TSU is required to focus on its R&D, and advance the technological aspects, to make sure it increases the market and regulation understanding and incorporate them into the R&D process as to make is the more effective for the future.

Regarding the means that the above can be achieved, some of the support which the TSU lacks internally can instead be supplied from external sources. Such sources can either be sources the TSU pays for in cash or equity, but in many cases, they are sources that TSU can access in a healthy SU supportive ecosystem. Starting with presentations of standard issues, or consultants' presentations as a marketing step for the consultants to groups of TSU's, accountant and different other experts, to downright mentoring sometimes of different voluntary organizations, government entities etc. are all supportive for the TSU.

The creation of such an environment and making sure the TSUs have access to it, are part of the important support any organization that wishes to see a growing TSU sector must support. Funding alone will not suffice, the human aspect is most important as well as the creation of the open culture that would allow the required access[8] (Jiménez and Sanz Valle, 2011). We have mentioned above, the need of the entrepreneur to learn all the time. This requires access to resources as well as time.

In that regard, the prevalence of places offering real-estate for SUs, has created several models each offering different levels of support and environment. Some offer just the place with the communication infra-structure, while others offer mentoring, courses and more intense support.

It is important to recall for the SUs that, they regard each other very much as co-*competitors* as Brawn termed it (Brown and Eisenhardt, 1998), meaning that they are both competitors (certainly for funding and other resources), but also cooperators, in the struggle to survive, learn and adapt in the non-technical aspects of their work. This cooperation is the basis of their community which allows them to participate in the ecosystem.

Practical Aspects

The above may seem to be a bit too general, and therefore some examples as well as some emphasis would be presented here.

One of the important aspects of entrepreneurship is the need for constant updating. As mentioned above, education and information for the entrepreneur (Barr, et all, 1992) are most important. Especially in the early stages where the team is small and limited in the range of expertise it can present, the information and education are important. In a mature firm, where experienced workers perform their tasks and the information channels internally and externally are established, the need is fragmented (logistics does not require updates related to finance etc.) and is handled by experts in their fields, and thus the need is less critical. Also, while efficiency is important as mentioned above, mistakes rarely are of critical nature. In the TSU things are different. As the small team would from time to time have to operate outside their field of expertise, updates are important, mistakes can be critical.

Thus, a major issue of support for the TSU which the ecosystem would need to supply on regular basis is information.

Another complementing, support measure, is the support of expert services. These can range from accounting and legal advice, marketing, finance, logistics, patenting etc., to complement the mostly internal services different departments in a mature company would supply, or at least would be responsible to out-source. The availability of such services as an out-source is important but the TSU team would still need to be able to explain the required service, select between options and understand the implication. If the team is updating a business plan and hires a marketing expert to improve the marketing section, it needs to be able to discuss and understand what the expert refers to as “market share”, “market segment” and know the difference. This may seem trivial, but that related to all areas that are outsourced, and one needs to remember that the TSU team would normally be technologically oriented (e.g., engineers) and probably not all that familiar with all the relevant disciplines.

These aspects create the built-in advantage of experienced entrepreneurs and managers when referring to the success chances of a new venture. They know more, have more links in their own network to such experts and know how to work with them.

Thus, the ecosystem should have an education aspect, it should also offer experts in various fields, or the access to such experts – a network of service providers, but it should do more.

There are three critical resources common to all TSUs: Time, HR, and financial resources and their power to purchase other resources.

Regarding time, the ecosystem can help in a limited way by shortening the search time for out-sourced service providers for examples, by helping through accompanying experienced advisers to avoid time consuming mistakes etc., still in a limited way.

The importance of HR to the success of the TSU cannot be exaggerated. The experienced and multi-talented members of the team, the way they come together and solve problems and advance the work is critical to the TSU success (Schneider, 1998; Zhang and Jia, 2010). This is so universally acknowledged that there are formulae for issuing shares for team members in order to encourage them to join the TSU and share its fate. The emphasis is on experienced members, in all fields and domain of the work, that know what needs to be done and how, that have the right experience and the right network to achieve results, and not just make the effort. In the TSU this extends from general management and marketing capabilities to the technical fields as well. In certain regions with limited employment mobility, the availability of such people can be a limiting factor to the creation of an active and vibrant SU ecosystem[9].

Regarding financial resources, it is a simple issue to understand and a complex one to create. The TSU is always aiming for the next round of fund raising, as this is the TSU's input of resources and its lifeline. Therefore, if the ecosystem can offer meeting financing entities for TSUs, to create finance opportunities (to meet the financing entities) this would be a vital element in the ecosystem (Sarma, S. and Marszalek, J., 2019).

When dealing with TSU it is important to remember that there are different kinds of financing entities for TSUs and to create a vital and energetic ecosystem it should offer the whole range. Angels or early investors, although often small, the amounts invested are vital from the very early stages, without such investors TSUs would find it extremely difficult to commence operation and the number of new TSUs starting every year would be small (Yiu, D., and Lau, C., 2008; Sarma, S. and Marszalek, J., 2019). The VCs and other first and second round investors would be required to continue up the growth chain of the TSU all the way to early stock markets for start-ups or/ and a market for mergers and acquisitions (M&A) to allow the TSU to mature to the last stage.

The fact that the creation of such an ecosystem containing all the above-mentioned financing entities, is easier said than done, can be demonstrated by the efforts of the EU Commission to create such an ecosystem and to increase the access to VC funding for EU SMEs (EU, 2003; Cohen, 2006; Feld, 2012; Audretsch and Belitski, 2016). The main challenge in the creation of such a fauna of financing entities is that they need to have a large enough number of TSUs at the relevant stage for them, otherwise they need to find them elsewhere. So, if you need to TSUs to get the financing entities, and the financing entities to create the TSUs, you find yourself inside a Mobius Ring (EU, 2003; Ács et all, 2014; Asheim et all, 2011). The process of developing such an ecosystem is therefore long and needs to be done correctly to assure that there are no missing links in the financing chain, else the ecosystem will not be sustainable, and the TSUs will migrate to places where they can continue to get the finance (Lerner, 2000; Spigel and Harrison, 2018).

FUTURE RESEARCH DIRECTIONS

While research is ongoing regarding the need of the entrepreneurs in TSUs to continue studying and the changes the mode of study (insert source from the recommended literature), along the TSU life. There is wide on-going research into the formation of TSUs, their maturation and the different activities accompanying this process (Spigel and Harrison, 2018). On the practical side there is on-going effort from practitioners as well as academics to understand the development and impact of the ecosystem, in order to further the development of TSU supportive ecosystems – enhancing economic performance over all (EU, 2003; Ács et all, 2014; Asheim et all, 2011). Regarding issues related to the complete firm – we shall discuss those when discussing the complete firm itself.

CONCLUSION

In this chapter, we have covered several aspects relating to the TSU. We started by exploring the nature of the TSU, the special characteristics and the demands placed on it.

We have mentioned the organizational and cultural characteristics, including the managerial tools relevant to the specific open culture of the TSU.

The specific characteristics regarding HR, recruitment, hierarchy, and job descriptions derived from the TSU culture. We mentioned flat structure and simple hierarchy of STU; the special unclear job description; the STU culture of openness and involvement in all. Following that, we discussed the different aspects of the TSU management. Its demand on the founders, and the resulting impact that the move to a complete company, will have on the management by the founders, including their decision to stay or leave.

We discussed the fit to the R&D phase, and the special importance of the ecosystem, the special support it can give the TSU, and the special needs of the TSU, and its dependence on external support. We mentioned the cooperative aspects of the TSUs in the ecosystem. The knowledge and learning demands on the entrepreneurs, and some examples were given. The scene is set for a description of some other forms of new ventures, to better highlight the special characteristics of the TSU, in the next chapter.

REFERENCES

Ács, Z. J., Autio, E., & Szerb, L. (2014). National Systems of Entrepreneurship: Measurement Issues and Policy Implications. *Research Policy*, *43*(3), 476–494. doi:10.1016/j.respol.2013.08.016

Almeida, F., & Miguel-Oliveira, J. (2022). The Role of Intrapreneurship in Portuguese Startups. *Periodica Polytechnica Social and Management Sciences*, *30*(1), 70–79. doi:10.3311/PPso.17796

Asheim, B. T., Smith, H. L., & Oughton, C. (2011). Regional Innovation Systems: Theory, Empirics and Policy. *Regional Studies*, *45*(7), 875–891. doi:10.1080/00343404.2011.596701

Audretsch, D. B., & Belitski, M. (2016). Entrepreneurial Ecosystems in Cities: Establishing the Framework Conditions. *The Journal of Technology Transfer*, *42*(5), 1030–1051. doi:10.100710961-016-9473-8

Barr, P. S., Stimpert, J. L., & Huff, A. S. (1992). Cognitive change, strategic action and organizational renewal. *Strategic Management Journal*, *13*(Special Issue), 15–36. doi:10.1002mj.4250131004

Brown, S. I., & Eisenhardt, K. M. (1998). *Competing on the Edge*. Harvard Business School Press.

Brunkhorst, H. (2020). "Everything is now in order" – Groupthink, ideology and practical-critical activity in the European financial crisis and beyond. *Journal of European Public Policy*, *27*(9), 1406–1423. doi:10.1080/13501763.2020.1773515

Calantone, R. J., Cavusgil, S. T., & Zhao, Y. (2002). Learning Orientation, Firm Innovation Capability, and Firm Performance. *Industrial Marketing Management*, *31*(6), 515–524. doi:10.1016/S0019-8501(01)00203-6

Cohen, B. (2006). Sustainable Valley Entrepreneurial Ecosystems. *Business Strategy and the Environment*, *15*(1), 1–14. doi:10.1002/bse.428

Denison, D. R. (1990). Corporate Culture and Organizational Effectiveness. New York: John Wiley & Sons.

Denison, D. R., Janovics, J., & Young, J. (2005). *Diagnosing Organizational Cultures: Validating a Model and Method-Working Paper*. Denison Consulting.

Entrialgo, M., & Iglesias, V. (2017). Are the Intentions to Entrepreneurship of Men and Women Shaped Differently? The Impact of Entrepreneurial Role-Model Exposure and Entrepreneurship Education. *Entrepreneurship Research Journal*, *20170013*(1), 1–14. doi:10.1515/erj-2017-0013

EU. (2003). *Communication from the Commission; Investing in research: An action plan for Europe*. COM (2003) 226 final. https://ec.europa.eu/invest-in-research/pdf/226/en.pdf

Feld, B. (2012). *Startup Communities: Building an Entrepreneurial Ecosystem in Your City*. Wiley. doi:10.1002/9781119204459

Gabai, U., Ehrlich, E., Abdel Razek, K., & Krieger, N. (2020). *High-Tech Human Capital Report 2019*. Israel Innovation Authority and Start-Up Nation Cantral.

Groupthink, B. (2013). Collective Delusions in Organizations and Markets. *The Review of Economic Studies*, *80*(2), 429–462.

Hasliza, A. (2014). The Growth of Innovative Performance among SMEs: Leveraging on Organisational Culture and Innovative Human Capital. *Journal of Small Business and Entrepreneurship Development*, *2*(1), 107–125.

Jiménez, D. J., & Sanz-Valle, R. (2011). Innovation, Organizational Learning, and Performance. *Journal of Business Research*, *64*(4), 408–417. doi:10.1016/j.jbusres.2010.09.010

Kang, H. D. (2018). A Start-Up's R&D Stages and the Evolution of Financing Sources: Evidence from the Biotechnology Industry. *Entrepreneurship Research Journal*, *20170159*(3), 1–19. doi:10.1515/erj-2017-0159

Korbet, R., & Analyst, M. (2019). *The State of the Israeli Ecosystem in 2018*. Start-Up Nation Central.

Lerner, J. (2000). The government as venture capitalist: The long-run impact of the SBIR program. *Journal of Private Equity*, *3*(2), 55–78. doi:10.3905/jpe.2000.319960

Mustafa, M., Lundmark, E., & Ramos, H. M. (2016). Untangling the Relationship between Human Resource Management and Corporate Entrepreneurship: The Mediating Effect of Middle Managers' Knowledge Sharing. *Entrepreneurship Research Journal*, *6*(3), 273–295. doi:10.1515/erj-2015-0004

Noy, E., & Ellis, S. (2003a). Risk: A neglected component of strategy formulation. *Journal of Managerial Psychology*, *18*(7), 691–707. doi:10.1108/02683940310502395

Noy, E., & Ellis, S. (2003b). Corporate Risk Strategy: Does it Vary Across Business Activities? *European Management Journal*, *21*(1), 119–128. doi:10.1016/S0263-2373(02)00159-7

Porath, A., Ramos, I., & Rahman, H. (2015). Different approaches to technology transfer by government and academy. *World Review of Entrepreneurship, Management and Sustainable Development*, *11*(2/3), 296–307. doi:10.1504/WREMSD.2015.068585

Puri, M., & Zarutskie, R. (2012). On the Life Cycle Dynamics of Venture-Capital and Nonventure-Capital Backed Companies. *The Journal of Finance*, *67*(6), 2247–2293. doi:10.1111/j.1540-6261.2012.01786.x

Sarma, S., & Marszalek, J. (2019). New Venture Growth: Role of Ecosystem Elements and Prior Experience. *Entrepreneurship Research Journal, 10*(1). . doi:10.1515/erj-2018-0215

Schneider, S. C. (1998). National Vs. Corporate Culture: Implications for Human Resource Management. *Human Resource Management*, *27*(2), 231–246. doi:10.1002/hrm.3930270207

Shiller, R. J. (2017). Narrative Economics. *The American Economic Review*, *107*(4), 967–1004. doi:10.1257/aer.107.4.967

Spigel, B., & Harrison, R. (2018). Toward a Process Theory of Entrepreneurial Ecosystems. *Strategic Entrepreneurship Journal*, *12*(1), 151–168. doi:10.1002ej.1268

Tsegay, B. K., Shi, C. S., & Wu, W. W. (2016). The Impact of Innovation Strategy on Organizational Learning and Innovation Performance: Do Firm Size and Ownership Type Make a Difference? *South African Journal of Industrial Engineering*, *27*(1), 125–136.

Yin, B., & Lou, J. (2018). How Do Accelerators Select Startups? Shifting Decision Criteria Across Stages. *IEEE Transactions on Engineering Management*, *65*(4), 574–589. doi:10.1109/TEM.2018.2791501

Yiu, D., & Lau, C. (2008). Corporate Entrepreneurship as Resource Capital Configuration in Emerging Market Firms. *Entrepreneurship Theory and Practice*, *32*(1), 37–57. doi:10.1111/j.1540-6520.2007.00215.x

Zhang, Z., & Jia, M. (2010). Using Social Exchange Theory to Predict the Effects of High-Performance Human Resource Practices on Corporate Entrepreneurship: Evidence from China. *Human Resource Management*, *49*(4), 743–765. doi:10.1002/hrm.20378

KEY TERMS AND DEFINITIONS

Advisory Board (AD): The AD is an organ of the firm that can be used to advise the manager on different issues where decisions or the ability to reach new contacts are required.

Complete Firm: This is a for profit business-oriented company, economically supported mainly by the income generated from its business (can also be supported by investments). This is what we normally refer to as firms, or companies. It has the full complement of activities and the managerial ability to manage them required to generate the income it needs to sustain itself and its operation potentially forever.

Ecosystem: The economic, social, and regulatory environment that the TSU come into interactions with. This includes funding sources, employment, and social laws, as well as other relevant regulatory measures pertaining to the activity and livelihood of the TSU.

Start-Up (SU): A general reference to a new enterprise whether for profit or not. It is mostly used for new companies or businesses formed and referrers to the early stages of their lives.

Technological Start-Up (TSU): A special case of the SU characterized by being based mostly on technology and preoccupied by performing mainly technological R&D and being economically supported mainly by investment and not by income generated from business activity. In may researches and publications it is this type that is referred to as Start-Up. For our discussion it is important to differentiate between a TSU and a general SU which can be a social or business-oriented entity not dealing with technology or R&D at all.

ENDNOTES

1 The Israeli tax laws defines such a company as a company where the majority of its activity is R&D. https://www.btl.gov.il/Laws1/02_0103_100136.pdf Hebrew.

2 Technology Readiness Level downloaded 5/5/19 from https://ec.europa.eu/research/participants/data/ref/h2020/wp/2014_2015/annexes/h2020-wp1415-annex-g-trl_en.pdf

3 (Gender) – the reader may have noticed that while I have tried to keep the discussion gender neutral, when required I refer to the TSU entrepreneur as male (Entrialgo and Iglesias, 2017). The only justification is that the majority of technical entrepreneurs are male. However, the discussion is relevant for all genders.

4 We shall deal with the definition of a successful TSU later on.

5 Just to allow for economically inclined readers, the market does not define the price but rather the demand / supply curve for the products / services. For the purpose of easy discussion, I shall continue to treat it as price setting.

6 Classical definition, real life advantages can be local, temporary or virtual.

7 Other possibilities than those in the next section are discussed in other sections in this book.

8 One of the attracting aspects of WeWork is this additional support it offers the clients, from services offered by the other residents, and up to external lectures and courses it offers (https://www.wework.com/en-GB/ideas/cancer-survivor-tapped-in to-the-wework-network).

9 Recent survey published in Hebrew regarding Be'er Sheva growth engines download from: http://www.amiramporath.co.il

Chapter 3
Other New Firm Types

ABSTRACT

Before moving to the comparison of the TSU with the complete firm in order to understand the challenges the TSU faces in the transformation to a complete firm, a short review is given in this chapter of other entities similar yet different from the TSU to complete the scene so to speak. The social SU and the bootstrap are discussed, and the differences between them and the TSU are explained. They are not without challenges of their own, and some of those are described in this chapter. The non-technical SU is also presented and discussed. The discussion is aided by some graphic work intended to clarify the comparison points between the different types described. Once the "fauna" is more comprehensively presented, the discussion is ready to move to the complete firm in the next chapter.

DOI: 10.4018/978-1-7998-5685-6.ch003

INTRODUCTION

The TSU is not the only new entity format for establishing new activities. This chapter will describe shortly some of the others in order to complete the picture for the reader. Business start-ups will be discussed showing that they differ mostly in the fact that they require from day one a complete firm, and that they therefore while having other problems will rarely face the same difficulties moving from the SU stage to the mature stage – albeit they will have their own difficulties. The analysis will start with non-business-like SU, progress to discussing business SUs without R&D, and end with a discussion of a technological new firm that is not a SU, the "Boot-Strap" model.

Before discussing the different types, it is important to remember that the different types have to play the ecosystem game as well as the TSU, and the Complete Firm to be discussed below. That is to say, that they play in their own ecosystem in which they have to adapt to a certain environment, and that they, or other interested parties, have to create that ecosystem in order to let the entities flourish (Danison at all, 2005; Audretsch and Belitski, 2016; Spigel and Harrison, 2018).

There are also discussions regarding gender and the impact it has on the selection of types and activities in innovation. This will not be further discussed here in order to keep the discussion focused (Entrialgo and Iglesias, 2017). Similarly we shall not discuss here risk preferences of the entrepreneurs (Noy and Ellis, 2003a; 2003b).

In the discussion some of the characteristics will be described with examples to demonstrate the meaning of each. The similarities and the difference compared with the TSU will be presented. It is not possible to describe each model to the full in such a short space so generalization will be used, which will of course drive to some inaccuracy, but will contribute to the effectiveness of the discussion. The main aim is to highlight some of the characterization and growth aspects of the TSU by contradicting them with other models.

The Social Start-Up

The social SU is not a business, but it comes to fill a social function, either taken up by the government and neglected or in answering a new social need. The social SU will normally provide a service and as it will not be business oriented it will raise funds in ways relevant for it. While depending on the local legal system for finding the resources it requires, funding, volunteers etc., it will normally have all the required functions from day one and will grow in size as required horizontally – inside each department.

Governments have been getting out of social functions slowly over the last two decades, and the process is still going on (Schneider, 1998; EU, 2003). The exit of the government from different social functions, combined with an increase in the number and diversity of social needs has led to the creation of many social enterprises.

The place of the government in taking care of social needs as well as identifying the needs is being filled by legal entities empowered to assist civilians that care about a specific social issue. The legal entities are in most cases associations or other non-for-profit legal entities, depending on the local legal system. These organizations raise funds and recruit other resources, such as volunteers and other workers, in order to facilitate their social activity and deal with what they regard as a major social issue. In most cases the activities to raise the funds – fund raising events, sales, sponsored events etc., as a facilitator for the main activity of the organization.

An example evident for a long time for such an activity, and not recognized as such by many resources is the academic research university, which raises funds in addition to tuition and national support from philanthropic donators, alumnae and other sources to sponsor side activities of the organization, such as additional buildings for teaching or research, dormitories and similar.

As this type of organization is better known, as is its philanthropic fundraising, it could serve as a good example for some of the characteristics we shall discuss below. At first look it would seem that the academic organization is raising funds for the general benefit of society.

The funds for research and teaching buildings, as well as for dormitories aim at the improvement of teaching conditions, and at allowing the young people attending these places to get the education required for the continual operation of society. In the age of knowledge and innovation, education is a major social asset.

However, these actions being supported, the buildings, the infrastructure etc. are not creating any new revenues for the organization. There is no new income derived from their operation, as befits a non-for-profit organization. This means that the organization must find new funding sources every year to fund its special social activities.

These two main points, that the operation of the social action requires specific fund raising, and that the activities do not generate the resources require to carry them on, are going to be major discussion points hereafter.

The social enterprise, in general does not differ in the need to start the process and to begin operation from a commercially based operation. In most cases it would also require an economic basis for its operation. It would require a client base, workers (volunteers or employees) and other resources, and supporting ecosystem.

While the business SU is focused on generating income and thus deals with the problems of offering the service / product, generating revenues, and creating a client base – hold on the market, the social SU has slightly different but also varied worries.

The social SU is formed to solve a social problem, either one that was taken care of by authorities (local or national / federal) or newly created and thus probably only partially recognized as general problems.

For example, an association looking to provide elderly lonely citizens with warm cloths, heating and food during cold periods, is offering a service that was formerly offered by the state, or municipality, and one that is recognized as a problem that needs to be taken care off. An association that is looking to provide guidance to the elderly in using social networks to keep in touch with their distant relatives, is providing a service that probably was never supplied before, and that is probably not recognized as a real problem by many.

Both of the above however, face problems related to the recruitment of resources for their activities, legitimizing their actions and goals. As Michael Porter mentioned in his TED talk (Porter, 2013), SU's face the unique problems, that unlike businesses where the goal is to produce the resources for continuous activity, the social entities' main goal is an activity devoid of resources creation, and thus they need to perform the activity for which they were formed and at the same time worry about their resources.

The main need for a social SU is to generate legitimacy. Without the legitimacy, the social SU cannot hope to be able to be formed, to get the resources it needs to operate and to perform the tasks it wants to do.

That legitimization is often complex, comprising both a marketing effort and a regulation challenge. The social SU therefore has to be able to contend with administrative issues, regulation, legal and social dogma and overcome all these obstacles. While it may be easier to explain the legitimate need when replacing the government, it sometimes is more difficult as you need to compete with the government standard of the past.

Once legitimization is obtained and there is a "social license" to proceed, the next step is to make sure that the resources required for the operation of the social SU are in place. These can be divided into two categories: HR and other tangible resources (including funding).

The HR Challenge of the Social SU

It is important to remember that many social SUs relay on volunteers as well as employee to a certain degree. That mix presents a managing challenge from recruitment to HR management and all the way up to termination of activities.

The recruitment of the employees is similar to that of a business SU, and in some cases a bit easier as technical or educational demands may be slacker, as the jobs do not require technical skill in most cases. However, some difficulties in recruitment may be the result of the lack of resources prevalent for most social SUs.

The recruitment of volunteers relies both on the legitimization of the actions, and goals of the social SU, as well as its reputation and marketing abilities. The social SU is required therefore to invest in promoting its own marketing image and to make sure the relevant target audiences (stakeholders), are reached.

A legitimate social SU, with a well-defined goal and a reputation for striving to further the goal socially, would find recruiting volunteers easier than a newly formed social SU, with a legitimate goal, but without a reputation of achievement (Zhang, and Jia., 2010).

The discussion of the management of a mix of employees (salaried HR) and volunteers merits a volume by itself. Here we shall state only that the mix requires agility on the manager side, the ability to motivate an achievement-oriented group, while also maintaining a good level of motivation in the parallel group of effort motivated people. Neglecting either can be very problematic.

Termination, which at first sight may seem to be the easiest is deceptively so. Termination of the employment of a single person either as employee or volunteer is normally not a real problem. But termination as a result of the termination of the activities of social SU. Here, as such termination is often the result of lack of resources, the cost of termination is a critical issue.

The termination of the employees is normally structured within the relevant employment laws and in the employment contracts. The employment of the volunteers, especially intensive employment, or long-term employment of volunteers is not so well regulated in most countries, which leads in most cases, to either law-suites, delaying shutting down operation, or in many cases in overcompensation as regarding the minimum required, in order to avoid the law-suites.

The management of the two kinds of HR is made easier on all three phases mentioned above, the clearer the difference in their tasks is. If it is made clear, that all office work is carried by employees, including administration, regulation etc., and that all external activities, such as working with the social cause, is handled by the volunteers, then it is easier to manage each group according to its nature, and to terminate each group separately. If both groups perform similar tasks, and in the same places, it would be more difficult to explain the difference.

Summary of the Social SU

The biggest challenge for the SU is the need to generate income and recruit resources separately from the main social activity it was created to perform. There is of course a question of legitimacy which states that a legitimate SU performing its social task would be able to generate enough public support, from the general public or private sources, to raise future funding. However, it is, unlike a TSU or business organization, a separate task from the social activity. This dis-connectedness between the social activities for which the organization was founded, and the resource generating activities that makes the Social SU a unique case.

The Social SU while being an economic organization as it requires resources to fund it operations, is non the less, a non-for-profit organization, and this dichotomy between the need to be economically oriented and keeping in mind that it is a non-for-profit organization, that makes the management of such organization a big challenge. Another one is of course the plethora of such organizations competing on the public willingness to be philanthropic (Jaskyte, 2004).

The challenges that such organizations face when trying to grow are mostly logistic wise. How to raise more resources, how to organize more activities or activities over a bigger domain etc., they are required for expanding activities similar to existing ones, and thus the structure and type of professions involved are normally the same and the managerial challenge is that of scope rather than multi-disciplinary. That is no mean challenge, but it is different in nature than the TSU challenge we discuss in this book. Especially as in most cases the manager comes from a discipline related to the social function, and not the fund-raising management training.

Non-R&D Start-Up Types and Characteristics

There are business-oriented-SUs, that are not technology based, do not require an R&D activity, such as a travel agency, an internet shop, a restaurant etc. They are different from the TSU fundamentally. They are a business from day one and like the social SU will require the full complement of activities to be performed in order to survive. Their growth will normally be more in numbers and rarely in branching into new activities. The differences will be described here and analyzed.

Most of the SU's that are not R&D dependent are also offering services and not necessarily a product, even if production is involved. Consider for example a new restaurant, there are products – the food, dishes and meals that are offered, but that activity may be considered a service. Providing meals and more for the clients. But even if they do sell products, such as a shop, they are non- R&D dependent. Market surveys are not considered for our discussion here as R&D.

Following the preamble above, in this section we deal with SUs that mostly offer service, even if a product is included, that do not perform R&D as a major activity, but may deal with experiments, again the new dishes in the restaurant would entail some experimentation but cannot really be considered as an R&D activity.

What further characterizes these businesses is the drive to produce income from activation, and hopefully to generate profit asap. These business enterprises may have some delay in producing the revenues – consider a consulting service that cannot collect until the consulting is finished, and longer to produce the profit, and this may be considered their SU period.

They are, however, active in the market, seeking clients, delivering the services and seeking financing and financial management. There are other additional actions typical to the sector they are active in, but all behave from day one as a more or less "complete" business, albeit perhaps on smaller scale than in a later stage.

In the SU phase such businesses would try and reduce expenses as much as possible in areas that are not directly linked to the production of income. So, while the restaurant mentioned above may not save on the quality of the food purchased for preparation in it, or on publicity and marketing, it will more than likely try to save on its accounting and legal expenses. A common mistake would be to go to the higher end of the service providers in unrequired services, as a general policy of the business. A policy of "we use only the best" would be smart regarding the branding of the restaurant when dealing with the food stuff, cooking, design, service etc., but redundant when dealing with the design of the invoices, legal and accounting services etc.

Making the distinction requires experience, and saving in the wrong place, such as on the personnel providing the service could entail a high price. The drive to save on non-revenue generating activities is intended to increase the breathing time of the new service, based on the initial capital it has. One has to recall that these businesses in most cases are required to supply income to the investors, entrepreneurs, and that therefore, the duration until they provide such income is limited.

Another aspect is the control of the required resources. The business SU may deem it strategically more beneficial not to own all the resources required, but rather to outsource them or to charter them from other businesses, until it can bring about the benefit of size aspect, or in some cases, just afford to invest in owning them.

Consider a new business of catering for events. The entrepreneurs may deem it better to hire the dishes for serving from other businesses, or using outsourcing do the laundry, than purchases the dishes and own the laundry facilities.

For a larger, more mature business, which is looking to reduce the laundry costs, it may be beneficial to invest on owning and operating such an activity. That is of course not the case for a new starting business, still focused on creating a reputation and generating enough events to create a large enough income to continue operation.

It is important to note that these activities exist and are part of the operation of the business SU but are not owned by it. That is different from the TSU in which in the R&D phase, marketing, finance, and customer service may not exist, internally or externally.

So, the business SU is a small replication of the complete business, presenting all the activities of the complete organization focused on creating the market presence as well as generating income and profit. While some of the above mentioned activities related to the complete company may be outsourced or chartered rather than owned and fully controlled, the business is entire in concept if not in actual operation. The maturation process would be in this case would be to embed as many faculties and own them, as fits the size of the business. It is important to note that the growth we discuss here is not just an increase in size. It is not just hiring more people into the departments or promoting people from a single person operation into unit managers. This may happen but it is not indicative of the maturation process of the business SU.

At this early stage the growth means introducing internally new activities and new units, with sometimes new type of workers in new disciplines that may also require some organization re-organization. In the above mentioned laundry example, the introduction of the laundry activity as an internal activity in the business SU, will entail a change in the responsibilities of the operation manager, the creation of an even support unit which will take care of all storage and preparation equipment for the event and thus of a new organizational structure, with new definitions of the roles and responsibilities of the manager. In that aspect there is some similarity in the growth process and impact to the TSU as well shall see. Growing that business is a challenge, as it requires competing in the market from an under-dog position and growing in spite of competition.

The "Bootstrap" type

A technology oriented new business different from the TSU in that it is not based on an R&D activity but rather on a business making income from day one. Similar to the business SUs it is a full scope company, a complete company, as much as it size will allow it, and differs from the R&D focused TSU. The Bootstrap may also perform R&D, but it will rely on its own income to sustain that activity. The Bootstrap will have however one major advantage over the TSU – it will be a player in the market, know the market best, and will have an established reputation as well as a proven ability to make money. A mention of a mixed model will be made and explained.

The concept of the Bootstrap actually predates the concept of the start-up and certainly that of the TSU[1]. The fact that the Bootstrap (BS) needs to be able to sustain itself from day one, affects the inherent capabilities, the structure and the culture of the firm. It also affects some of the competitive advantages it would have compared to the TSU. The discussion in this section refers to a BS that later develops its own product / service to enter the market, and is, in this form, an alternative to the TSU. A BS that will remain as a service, even technical service over time is just a small service company, as mentioned in the section above.

The main advantages that the BS would have over the TSU are the following:

1. It is a player in the market. This fact leads to several advantage points:

a. It knows the market, has the ability to perform a market survey all the time, real time, and therefore have the ability to verify that whatever the BS develops, product, service, or both, answers a real demand.
b. It can accumulate a reputation and thus ease the entry to the market. The BS can use its market activity to establish the entry point of the product, weather it is a beta site at a leading customer, or connection to marketing channels best fitting the market segment it wants to enter.
c. The BS can find suppliers, sub-contractors, and HR that can best fit its need and test them.

2. It is self-financing and thus accumulates experience in addition to R&D performance and R&D management, in marketing, finance, HR, production sales and post sales as well as general management. That set of capabilities allows the BS to enter the market better prepared for the challenges such a change presents to the company. The accumulated experience allows the BS to prepare mentally, and culturally better for the market change that will appear if it enters the market with a new product. Even if it is not fully self-financing, even partial self- financing is a big step towards getting the rest of the required funding for the development, either as equity or as debt.

The setting of a BS is not easier than that of a TSU, depending on the ecosystem. In ecosystems rich in VC and angel funding, a TSU forming can be easier than that of a BS, as the funding and normally also the HR and managers are available over time. In an environment that is lacking in VC funding there often is no choice but to form a BS in order to start the business.

The BS for our definition, is a firm that is formed based on a small investment that is aimed to finance the early stages of the forming. The firm generates income from the first day, and it therefore needs to have the capabilities to generate the income. The BS can be formed first as a firm either selling a service or a simple product that does not require a long development stretch.

An example would be a small firm established by several computer experts. The firm starts with supplying consulting services to several large companies as an out-source. The BS starts to generate income, from which a part is kept employing several workers in the firm, that would work on developing the products that the BS is intending for the future growth of the BS.

The BS is active in the market, using the information it generates on daily basis, on the needs of the potential clients, to improve the characteristics of the future products. The BS has the chance to develop the product and test it with the assistance of some of the potential clients. It can also in that case decide to release an incomplete early version, or a beta-site. It has also an advantage in the ability to decide that it would like to stop the development effort, without risking the well-being of the firm.

The TSU ability to continue to operate, depends on its continued development effort. A TSU that received an investment cannot decide to stop the development of the promised product and expect the investors to continue the investment. In fact, not only would the investors not continue to invest, but they would also probably act to dissolve the firm and try and recuperate the funds already invested. The BS would not need to fear being dissolved, due to changes in the development, as its income is independent, of its development progress. In fact, one would expect that the available income of the BS would increase if the development stopped and the HR and other resources involved in development, would be directed towards generating more income.

The BS has a better experience in the management of the firm, including the ability to create for example a credit history that it can use to raise funds. The credit history of the BS places it on a different plain than the TSU. The BS, after a certain period can access dent as a regular firm and is not limited like the TSU to VC and plans for the future.

The greatest advantage of the BS is that it is a complete firm with full capabilities for generating resources based on its income. The BS does not need to go over a change in the more into the market, based on its development efforts. It needs to add sometime, additional capabilities, but the main effort is normally in scaling-up the efforts and not the changing the organizational culture.

The BS in the above-mentioned example is service based. Since the development effort is directed towards a new product, once the BS is ready to introduce the product to the market, it would need to add a production capability, but the sales and finance would require only adaptation.

If the BS is already based on a production of a product, the entry of the new product into the market would require even less adaptation. Culturally the BS may share characteristics with the TSU, such as the limited hierarchy and formality; the involvement of the managers in all BS aspects and job descriptions that are generally defined, where everybody does what is needed. Thus, the BS can also move towards maturity in these aspects.

The Summary of the Different Types and Their Characteristics

Presenting a summary of the above with their special characteristics, showing the advantages each has, and the structural and cultural as well as organization impact these have on the different types can be done in several ways. One way is presented in the drawing below - Figure 1 - SU types per technology and resources).

Figure 1. SU types per technology and resources

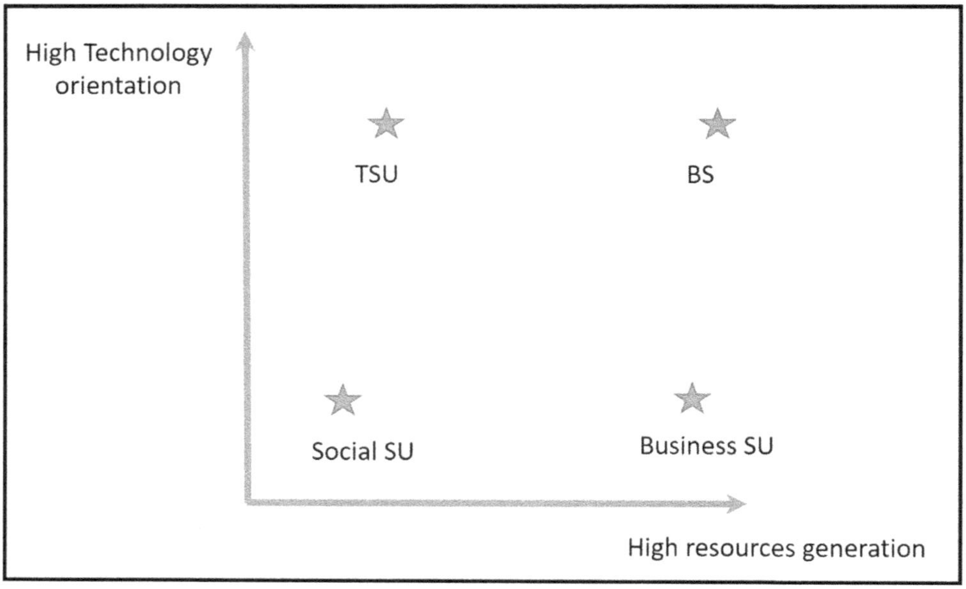

The different models mentioned and discussed briefly above highlight the characteristics of the TSU, as an incomplete firm with a special maturation process that includes not only growth in size and capabilities, but also entails the introduction of new activities, types of personnel and management requirements, as well as an on-going organizational structure change.

While the business SU and the Bootstrap both operate as a complete and mature firm, with income generation capabilities, some of the activities remain at the early external stages to the organization, either by out-sourcing or by chartering them. The introduction of these activities into the organization is a maturation process involving both growth and change and is derived by a combination of a wish for more control, and the drive for increased efficiency. The limiting factor is the economic ability to sustain them internally.

The social SU is a different entity as it has other aims than a for profit organization. There are many challenges when dealing with this type of entity, as we have discussed above. The important issue to remember that in the case of the social SU the elements of legitimacy and HR are crucial for its future survival.

While all types of SUs are characterized with similar challenges of starting a new operation, formulating the internal processes, acquiring legitimation and the resources required for their activity and survival, they differ in their character, their goals, and the environment they operate in.

Their growth challenges and processes are therefore also derived from the same basic challenges and the basic structure and type of activity. While some require mainly growth and assimilation of new employees, most require also structural changes and the incorporation of activities formerly chartered out.

The TSU will differ in this aspect. The TSU will not require exclusively growth and assimilation, or restructuring, to allow former out-sourced activities to be internalized. Rather, it is special in the need to contend with new activities, that it has never been required to perform, internally or externally, and that will challenge not only its structure, but mainly its management capabilities. This is in many cases a turning point in the TSU life, which even if it goes through a downsizing cycle later in its life, will rarely got back to the TSU structure.

It is in that respect, that the TSU is passing a threshold. Former rounds of fundraising, and growth, were similar to each other, but from this point, the nature of the organization has changed.

It is therefore also a time for the organization to undergo a major HR change.

The former employees, including in many cases the founders, would leave, with whatever gains they can, and probably found a new TSU, or be involved in related activities. While the rest and the newcomers, will find out that they are under different management with a different working environment.

Those leaving and founding a new TSU are going back to the familiar structure and culture we mentioned above. The others are members of a new type of firm, the "Complete Firm". But before we discuss in the following chapters, the change that they undergo, we need to discuss the "Complete Firm" and its characteristics in order to understand the change now required.

FUTURE RESEARCH DIRECTIONS

The research relevant for the different types discussed in this chapter deals with organizational as well as social theory and is therefore mostly out of the scope for this book.

The topic of social SUs, and especially their role in society compared to that of the government is a major research topic and as the advent of the social SU increases, so will the interest in issues relevant to it and thus the research of that topic.

The issue of Bootstraps and their formation and the non-technical SUs is part of the on-going economic research aiming to understand the parameters of economic growth, economic and social stability, and by reference also regarding governance the involvement of government in economic issues.

CONCLUSION

While this chapter may seem a bit out of focus for the discussion in this book it is intended to give a more comprehensive picture of the existing types of enterprises which may confuse the reader as they may be discussed in conjunction with the SU or TSU.

The SU as mentioned in the previous chapter being a generic name used for several different types of enterprises is split into several of the more prevalent types, and each is discussed and explained. The main types relevant to our discussion are the non-technical SU, which is characterized and the difference between it and the TSU is explained. Similarly, the Social SU is explained, and this also in reference to its role in society and the specific challenges this type of enterprise faces.

The main alternative to the TSU, the Bootstrap is presented and discussed, the discussion is relatively short as the nature of the Bootstrap dictates a different set of activities and thus a different set of managerial capabilities. That special set makes the special transformation discussed in this book less relevant for the Bootstrap and in turn, the Bootstrap less relevant for the discussion.

The reader is now ready to discuss the target of the transformation, the complete firm – the topic of the next chapter.

REFERENCES

Audretsch, D. B., & Belitski, M. (2016). Entrepreneurial Ecosystems in Cities: Establishing the Framework Conditions. *The Journal of Technology Transfer*, *42*(5), 1030–1051. doi:10.100710961-016-9473-8

Denison, D. R., Janovics, J., & Young, J. (2005). *Diagnosing Organizational Cultures: Validating a Model and Method-Working Paper*. Denison Consulting.

Entrialgo, M., & Iglesias, V. (2017). Are the Intentions to Entrepreneurship of Men and Women Shaped Differently? The Impact of Entrepreneurial Role-Model Exposure and Entrepreneurship Education. *Entrepreneurship Research Journal*, *20170013*(1), 1–14. doi:10.1515/erj-2017-0013

EU. (2003). *Communication from the Commission; Investing in research: An action plan for Europe*. COM (2003) 226 final. Available online at: https://ec.europa.eu/invest-in-research/pdf/226/en.pdf

Jaskyte, K. (2004). Transformational Leadership, Organizational Culture, and Innovativeness in Nonprofit Organizations. *Nonprofit Management & Leadership*, *15*(2), 153–168. doi:10.1002/nml.59

Noy, E., & Ellis, S. (2003a). Risk: A neglected component of strategy formulation. *Journal of Managerial Psychology*, *18*(7), 691–707. doi:10.1108/02683940310502395

Noy, E., & Ellis, S. (2003b). Corporate Risk Strategy: Does it Vary Across Business Activities? *European Management Journal*, *21*(1), 119–128. doi:10.1016/S0263-2373(02)00159-7

Porter, M. (2013). *The case for letting business solve social problems*. https://www.ted.com/talks/michael_porter_the_case_for_lettin g_business_solve_social_problems

Schneider, S. C. (1998). National Vs. Corporate Culture: Implications for Human Resource Management. *Human Resource Management*, *27*(2), 231–246. doi:10.1002/hrm.3930270207

Spigel, B., & Harrison, R. (2018). Toward a Process Theory of Entrepreneurial Ecosystems. *Strategic Entrepreneurship Journal*, *12*(1), 151–168. doi:10.1002ej.1268

Zhang, Z., & Jia, M. (2010). Using Social Exchange Theory to Predict the Effects of High-Performance Human Resource Practices on Corporate Entrepreneurship: Evidence from China. *Human Resource Management*, *49*(4), 743–765. doi:10.1002/hrm.20378

KEY TERMS AND DEFINITIONS

Bootstrap: A new technology-based enterprise with economic gain intentions basing its resources on economic gains from a commercial activity while endeavoring to perform R&D as the basis for future growth.

Non-Technical SU: A business-oriented SU without recourse to special or proprietary technology, (e.g., a restaurant or a shop).

Social Start-Up: A non-for-profit SU intended to forward some social goal, raising funds in order to finance its activities as these do not create the resources it needs for further sustenance.

ENDNOTE

1 Take for example the printing firm of Gutenberg, it could not be termed as a TSU, but rather as a BS.

Chapter 4
The "Complete" Firm

ABSTRACT

This chapter introduces in detail the complete firm and its characteristics, structure and culture aspects, and its challenges. The chapter completes the introduction of the different entities involved in the discussion of the book. It starts with the introduction of the enterprise called the "complete firm" and moves to characterize it, presents the organizational structure, and the culture developed in order to fit the demands of the environment the firm is active in. The management aspects of the complete firm are mentioned, and the challenge of managing each different aspect of it as well as the general management of all are presented. Lastly, as preparation for the next chapter, the TSU and the complete firm are contrasted regarding the focus of their activities.

DOI: 10.4018/978-1-7998-5685-6.ch004

INTRODUCTION

The discussion so far has focused on the TSU and its similar types of enterprises. In this chapter the discussion moves to present the Complete Firm, the next stage in the maturity of the enterprise.

The Complete Firm is an older type of enterprise to receive the focus of decision makers and of academic research and has thus a long history of relevant research (Penrose, 1959). This research relates to the many aspects of the firm, including the culture and structure of firms, and the resulting issues of management. It is a vast body of research and the source of numerous volumes of many pages. There is no intention to present here a comprehensive picture of a Complete Firm, or of the management of it.

In order to keep the discussion focused, the description keeps to aspects that were presented regarding the TSU, in order to emphasize the effort required for the transformation from TSU to a Complete Firm.

In order to understand that gap (to be discussed in the following chapter), the Complete Firm relevant aspects need to be presented, following by some comparison between the two.

What is the "Complete" Firm

A complete firm is a legal entity that performs the whole range of activities of a business, creating the resources it requires for continued operation independently in the long term.

Although we have discussed above some non-for-profit organizations, for the effectiveness of our discussion, the legal entities we shall be interested here are all business-like and profit oriented. Thus, the companies we discuss here are able to perform the entire range of activities that will allow them to survive in the long term, creating the profits they aim for and whenever possible attempting to grow.

Our vision of the complete firm does not mean that it runs the entire value chain internally. A complete firm is definitely one that outsources some of its activities, either temporarily or on a longer term, but still these activities are performed on its behalf, and consume to some degree some of the firm's management resources.

The firm size is not relevant to its completeness. It can be a small consultancy that still markets its services, provides them, and collects the reimbursement creating its own resources for the future. At the other end of the scope, it can be a multinational with hundreds of thousands of employees, affiliates, and sub-contractors, with numerous products selling them all over the world, and with the relevant departments. That firm would have a complex structure but may also have some of the activities outsourced. The main similarity lies not in the size or age, but for our discussion will relate to the managerial scope of control, and therefore the managerial capabilities requirements.

Just for clarification, we assume that even out-sourced activities require some managerial oversight and control, even if not daily management like internal activities, from the management team of the company out-sourcing.

An incomplete firm would be an R&D division of such a multinational mentioned above that deals only with R&D, transferring its results to other units in the company (even if it registered as a company, an independent legal entity), or a production division that was span off, and does not deal with marketing, with sales or any other activity except production. Of course, our main aim is to differentiate the TSU from the complete firm.

The discussion of the managerial team capabilities and the challenges that the complete firm faces, are a characteristic, but are not the only difference between a complete firm and an in-complete, or partial firm.

However, for our discussion in this book, we shall differentiate between the TSU and the complete firm, by these two parameters: the scope of managerial capabilities the management team faces; and the parameters for the organizational structure and the position definitions as the second parameter.

The size is not a clear indication, as there can be relatively small complete firms while next to them in the same market, we can find larger TSU's. It is true that size can increase both the magnitude and the variation in the managerial challenges management would face, but it is not a clear indicator. A TSU with 30 employees would still be mainly dealing with R&D with no marketing, sales or after sales activities, while a complete firm of 25 would face all these problems and have to deal with them.

The age of the firm is not a clear indication as well. Some TSUs trudge along for years until they manage to reach a breakthrough or move somehow to the next stage. While a business SU like we discussed above, or a Bootstrap would be dealing with the problems of a complete firm from day one. The move from one status to the other is again not a question of size or age, even though these could hint at the readiness of the firm to move from one stage to the other, but rather of the changes in the characterization of the jobs in the firm and the resulting impact on the organizational structure, and the managerial capabilities. There different ways of dividing the maturing process of the TSU into a complete firm, but most seem to struggle with the definition of the border line between the two. Some quote a certain revenue/ profit quality, others independence etc. (Santi and Neves, 2020), here we refer to a different differentiation.

Another misleading characterization is the rate of growth of the firm. TSUs, may from time-to-time experience rapid growth rates, which may indicate a critical stage change in the development process, but not necessarily the move from the TSU to the complete firm. On the other hand, the fact that the firm is growing at a moderate rate is also not a clear-cut indicator that it is a complete firm. TSUs experience periods of slow growth or even decline in size.

The complete firm is an organization that is able to sustain itself in the long run, based on its own income derived from its usual operation. The specific characteristics of the complete firm are the capitalistic nature of its action, its derive towards revenues and profit, and the fact that in the long run, it requires income in order to sustain itself.

Depending on the specific sector and they type of product, service or combination of both, the legal entity will be required to supply itself with the resources for its survival. It may from time to time raise additional funds, either per equity, or based on debt, but it will do that in parallel to its regular activities.

In some sectors it will be required to offer guidance and advising services, customer services, maintenance services, etc. and in others it will not be required to do so. However, the complete firm, in the bio-organic terms is a complete living creature able to sustain itself. If it is not able to do so, it will be terminated.

Therefore, the complete firm, due to its nature faces several challenges.

Challenges the "Complete" Firm Faces

Times are changing and competition increases with globalization. It is not easy for firms anywhere, but there are different problems for different types of firms to be faced. The specific needs and the way the answer is reflected in the structure of the complete firm. In this section we discuss the changes and challenges the complete firm faces.

Ever since Penrose wrote about the nature of the firm (1959), the discussion regarding the different organs in the firm and their fit for their challenges and the environment in which they operate has dominated the academic discussion regarding firms, management, and structure.

The complete firm is an independent entity, facing the survival challenges of operating in a certain environment, recruiting resources, creating its own internal process, and seeking long term survival and aiming for success.

While this may sound wonderful – it does for me, it means that the management of such a firm is required to handle challenges over a wide range of activities and is required to deal with them successfully enough so as to assure that the firm survives. When we discuss the TSU challenges we shall discuss this point further, however, at this point we shall agree that the main difference in the point of view of the complete firm versus the TSU – is the length of the survival aims. The complete firm looks for long term survival and growth, the TSU, at managing the move to the next stage, or at best, at surviving to arrive at the next stage, in the best form to continue to successful funding. Complete firms that their management fails to seek survival for long term, tend to find themselves facing different forms of crisis from time to time, and regardless of the size may be taken over by other entities.

The challenges the complete firm faces can be generalized into the following categories:

Solving simultaneously problems and facing multidisciplinary challenges – the need to face the complex relations between R&D, marketing, and sales, with financial issues and allocation of resources can demonstrate this point. The problems the firm faces are usually interlinked with wide ranging implications over the entire firm. Financial status may limit the number of people the firm may employee, which in turn will impact the allocation of HR resources and thus the ability of the firm to perform certain tasks. It may decide that it can manage better financially if some of the tasks are outsourced, which will impact the organizational structure, and may impact the revenues or the growth of the revenues. Management literature if full of discussions on how to solve the different problems, how to design solutions, and some of the solutions are just as cross-disciplinary in nature and firm wide, as the problems. The management is required to be able to identify the problem and define it, and then design and implement the solutions. Solving a technical problem by coating a component in gold may satisfy the R&D department, but the marketing experts would enlighten the management to the point that the product would no longer fit the low-cost markets in which the firm is active (Zahra, 1995; Zahra et all, 1999; Baum, et all, 2001).

Maximizing the overall performance in the short term while not damaging the future possibilities for growth of the firm. This point deals with the types of the solutions with which the firm can deal. That set of solutions it can be argued requires from time-to-time re-evaluation as to the long-term impact of certain activities. The decision to distribute dividends to the shareholders, may have a positive impact on the status of the firm in the stock-exchange, and may make it easier for it to raise funds in the future, but it can also harm the investment abilities of the firm in production marketing and R&D. I emphasize the long-term point of view here, as the this would be one of the differences we shall discuss when we get back to the TSU. Losing the long-term view, can lead to a loss of managerial capabilities which in time would cripple the firm and put its survival in danger.

As an example, for the last point consider the following example: company in the small plastic item production starts with a technology developed in a university and becomes a leader in the manufacture of disposable plastic food dishes. After several years of leading the market, the company moves from an R&D activity, dealing with new production process and different materials, to a small set of production process and an even smaller set of materials, focusing instead on the design and decoration of its products. It dismantles the R&D department, and the people working there are dismissed and move to other firms. After an additional ten years, there is no one in the firm with experience in R&D, R&D management, or related activities. The market opens to cheap imports from low production cost countries, and the firm sees that its market-share, and sales diminish from year to year. The management consider different solutions to the problems, such as purchasing new production equipment, streamline the products to reduce variation and cost and other measures. The measures are all within the scope of capabilities of the management team. R&D is not considered internally as the management does not have the managerial capabilities to deal with it. It has to learn to define the problems design the R&D project, seek the resources to implement it, and be ready to implement the results. But as it is not familiar with the process, the management would not consider R&D as a solution, unless an external source suggests it with the help to implement it.

There are some tools (support programs – Porath, 2017), that can help such companies get back into the R&D game. However, for such programs to be effective a financing element is not enough, an R&D management capability injected into the management team is the crucial element[1]. The SME Instrument program in the EU DG Research Horizon 2020 program, allows therefore for the selected projects to be assisted by innovation experts once they have gone through the screening process. It is such programs that realize that the funding is not enough for firms who are aware that they are not capable to manage the project, and that they require external assistance. The SME Instrument Program is aimed to assist firms in making an innovative move that will take them to the market. We shall later on discuss the move to the market from the R&D phase, but at this point it is important to remember that this assistance is not aimed just at new companies, but also at veteran companies making a new innovative move that will help them introduce something new to the market.

Getting back to the management challenges, it is the interlinks between the different actions, and thus the departments in the firm, that influence each other, and therefore requires that the management be able to understand the potential impact of each action, and manage the firm (Hughes and Morgan, 2007; Nason and Wiklund, (2018).

I am referring to the management team, as except for very small organizations, the management in a firm is not a single person show, more than one person is involved. Even in small firms, the manager may be responsible for the sales, marketing production and purchase, but will normally have at least finance and HR support from others, and in many cases other aspects of the firm's activities as well.

In a complete firm, where there is production, marketing, sales and post sales customer service, as well as supportive activities such as: finance, HR, logistics and R&D, some of the activities may be internal, some external or outsourced, but still they would be connected and impact each other, and require at least some management assets to manage them, and coordinate them with the rest of the activities. Even out-sourced activities require management assets as they are linked to other activities and either impact them directly or are linked via the shared resourced limitation.

The shared resourced limitation is the condition in which the entire firm resources are limited, even if they are vast in general terms, and therefore when allocating some resources to certain activities, they cannot simultaneously be allocated to another. To illustrate that, imagine a firm specializing in silver coating of small artifacts. The silver is a limited resource for the firm, and any silver allocated to coating picture frames, cannot be allocated to coating jewelry. Thus, when dealing with the mix of internal and external tasks, at least the financial resources are limited, and thus funds allocated for external tasks impact internal tasks by making the funding for the internal tasks less available. There can of course be more complicated relations. If the logistics are outsourced, the marketing and sales activities may be limited by the restrictions placed on the firm by the logistics supplier. If shipments are available to shops only once a week, the firm cannot commit to delivery in 4 days. In some cases, when the market (the competition in most cases) demands that in order to compete you must deliver in 4 days, this is a serious problem for the firm which the management needs to resolve.

These are not insurmountable difficulties and firms deal with them on a daily basis. Management literature deals with such issues on a regular basis[2], and the solutions a numerous and known. However, this does not diminish the need of the management to be able to deal with all of them. The ability to cope with such difficulties requires the management team to be varied in skills experience and discipline.

In addition to the demands on the management, there are administrative demands imposed on the entire firm. While even the smallest firm may rely on simple management regulations and stream-lined administration, with the complexity of the activities and with the demands for quality assurance, most firms need to regulate their decision making, track the production processes and the allocation of resources and other administrative activities.

These controls and documentations regulations are required due to the complex activity of the firm and are not required, at least not on the same level of detail from the simpler single activity firms (like the TSU) in many cases. In the cases where such requirements are placed upon the TSUs (e.g., Pharmaceuticals), they are less complex than for the mature firm as they are focus on a single activity in most cases (e.g., good laboratory practice).

Organizational Culture Characteristics of the "Complete" Firm

The organizational aspects of a complete firm are a main field of interest to those dealing in organizational theory (Barr et all, 1992.; Jaskyte, 2004; Delmar and Shane, 2006). We shall try and limit the discussion in this section to the main issues that point out the difference between the complete firm and the TSU.

The challenges faced by the management of the complete firm are in many cases mirrored in the structure of the firm. There would be the people dealing with the manufacture, those dealing with marketing sales and post sales, and other fields of activities. The structure of the firm will reflect the different activities and the personnel dealing with them.

The size of the complete firm will have an impact on the organizational structure, however, that impact is also a reflection of the managerial span of control and the management structure. In small complete firms, there will be people who may be responsible for more than one activity, sales and post sales may be dealt with by the same people, production and service, or maintenance the same etc., however, they will be dedicated to the task, and will have clear guidelines how to deal with each activity, and the main point is that the linkage between the tasks in based on disciplinary proximity of the tasks. The people will be recruited for the main task and be expected to assist in related tasks.

The structure is therefore organized according to tasks and grouped under similar or disciplinary tasks. It will be easier to follow logically, and intuitively and will be easily expanded if the firm growth. The separate tasks would become separate departments under the same divisional management. Even structural changes will be mostly done according to the disciplinary proximity.

It thus sounds that to be an HR executive in a complete firm is a straightforward job. Technical people in the technical department, each with a job description of what the person in that position will need to attend to, and thus what professional training or experience that person should have.

It is much easier to imagine the complete firm to be a medium to large one. Imagine a plastic producer with up to a hundred workers employed. This time complete with an R&D department (see Challenges the "complete" firm faces, above). It has a manufacturing department with several manufacturing units, a unit for maintenance and a liaison unit to other departments. It also has marketing sales and post sales customer service as different units under a single department and a general department for management comprising the office of the CEO, the HR, Finance and legal units all in one. While it may culturally be classified under several headings in the organizational theory classification of organizational structures, it's culture will be characterized generally as striving towards efficiency, aiming at increasing the income, reducing costs, maintaining market share and increasing the profits of the firm. It will seek efficiency and effectiveness in all aspects. It may be more or less open to internal innovation, ideas etc., but it will generally focus on efficiency and profit increase in order to assure its growth at best and at least it's survival.

The same generalization could be made for a much larger firm, hundred thousand employees and more, with thousands of products, or a 20-person firm dealing with less than 20 products. They will have the same push towards efficiency and profit increase drive. Their structure may differ, and the culture may differ between different units in the larger firm, but the main characteristic mentioned here will exist in the general firm culture.

It is important to pre-judge this tendency to increase profit as a negative trend in the management of the firm. As the profit presents the firm's ability to produce the resources it needs for long term survival it is crucial to understand its important. Firms that forget that important aspect tend to disappear, and other firms take their place.

It is therefore very important for our discussion to remember that specific set-o- mind. Every decision is ultimately judged as to its benefit for the firm. That benefit of the firm, in measurable terms in the ability to generate profits. True, one may scarify profits in the near term to secure more profit, or other profit, in the long term, but not for the sacrifice's sake alone. That fundamental point of view will characterize for our discussion the complete firm, versus the TSU. It will also be the key for discussion regarding the gap between the organizations.

However, before we discuss the gap between them, we still need to address the legitimate structure and organizational culture of the complete firm. The legitimization stems from the fit to the environment and the fit to the goals of the firm, as we shall see in the next section.

In fact, the efficient and rigid structure of the mature firm, and its resulting organizational culture, have created a specific problem for the mature companies, especially the larger more rigid ones, and that is the ability to innovate internally (Owens and Fernandez, 2014; Porath 2017). This is considered in many cases the current challenge the larger firms have to face and there are different solutions to it as well. While this is not a major topic of this book, it is nonetheless important to present the benefit the larger firms derive from the M&A with the TSU. The larger firms, with their increased efficiency, levels of decisions and more rigid trouble, tend to suffer from a reduced ability to innovate, (Owens and Fernandez, 2014; Korbet and Analyst, 2019). One of the major sources of innovation for the larger firms, is the purchase of innovative firms. That process can also be used to reach the growth goals of the larger firms.

The fit to the Needs of the Firm

When organization experts discuss "fit" they normally focus on the fit between the organizations' structure and culture, to the environment and challenges it faces. These include organizational as well as cultural aspects. In this section we want to include in the discussion the goals of the organization and the time scope it refers to.

Let us start by first defining the environment in which the complete firm is operating. The complete firm will normally be operating in a complex environment. Legal aspects which are derived from employing people, owning facilities, producing revenues and thus incurring taxes, and other legal aspects that may come into the scope of management of the firm. However, the legal environment is not the only one. There is the financial environment, be it local, national or world-wide. It will have impact on the firm as regarding resources, market ability to absorb its products etc.; once again, this is not the only influence on the market, as there are the competitors, the suppliers and customers, and all entities that are some of them together. There are issues related to regulation, depending on the specific sector of the firm, and the country or region in which it operates. And many more.

While again, some of the above may be outsourced, the firm will need to deal with all of them in order to survive and thus will have to be able to fit the environment in order to survive. The fit in most cases, even for large firms, will probably include a mix of internal units specializing in some activities, and external services to comply or answer the rest. Thus, in the plastic example from above the management department with its units will comply to the legal, financial and regulatory aspects of the environment; while the sales and marketing department, and production department, will be there for answer the fit with the market environment, the suppliers, clients, competitors and those that are both at the same time. It is important to recall again, that the size is not the determining factor here. Although it will influence the decision which activities to retain and which to outsource.

In order to better understand the issue of the fit please see the two pictures below. The first one (Figure 2- TSU's environment), describes the fit of the TSU to its environment; while the second (Figure 3- The Complete Firm's environment) one, does the same for the complete firm.

As can be seen from Figure 1- TSU's environment above, the environment that the TSU "sees" is very strongly controlled by the wishes of the different investors, and the need to grow fast and create more knowledge. One of the strong influences, and in certain areas, a barrier to the fast growth of the TSU, is the availability of human resources required. The scarcity of human resources is especially important in cases when either the country / region in which the TSU operates is weak in that regard (e.g., Train engineers in one of the Mediterranean islands), or if the general cost of the HR has been increased due to demand of other companies, especially large companies (e.g., AI quality controllers).

Figure 1. TSU's environment

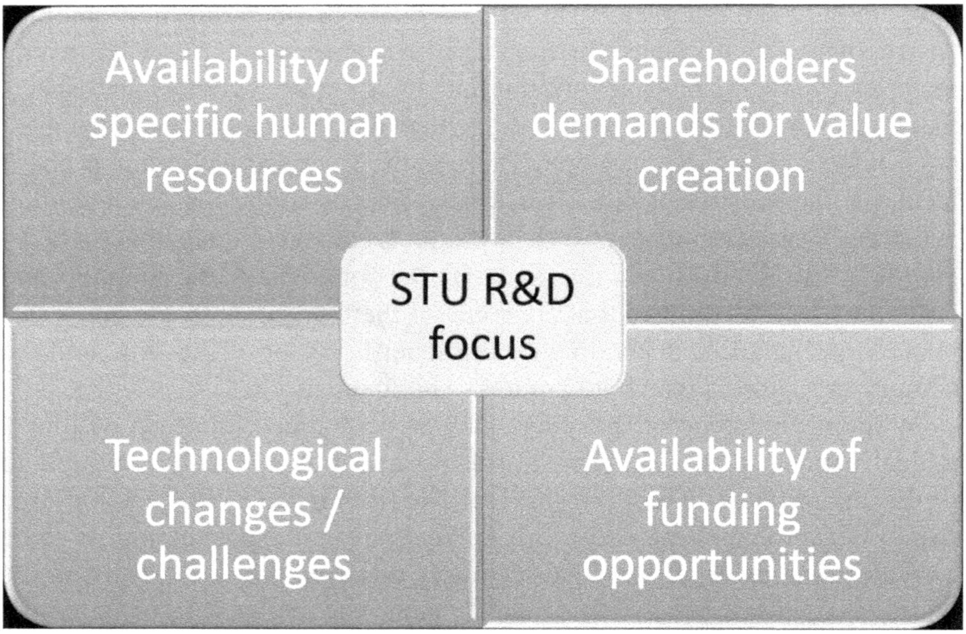

Another big influence either as a support for growth or a potential barrier is the technological environment, technological shifts and changes and other technological challenges. As the technical knowledge creation is the main value creation process in the TSU, any threats or potential changes to it, have to be closely monitored. The TSU will need to develop the tools to scan the technological environment it operates in and to identify threats as well as opportunities. The "Garage firm" is an extreme example in which few people are even aware of the existence of it, and thus it is not influencing the market at all. If your competition may come from there, you need to be able to identify it and analyze the threat.

Weaker influences, include among others, the market influence. While it is relatively weak, as the TSU, at least in the early stages, is more an extraneous observer than a player in the market. Regulation is a parameter in the R&D but again as it would in many cases be, either outsourced to external experts or deferred to a later time, the influence is secondary. Here the influence of the experience of the managers, as well as the investors, especially benture capital funds, in the firm can be seen (Gompers, 1995; Audretsch and Keilbach, 2004; Hughes and Morgan, 2007; Sarma and Marszalek, 2019).

In the second figure, Figure 2- The Complete Firm's environment we can see that the complete firm faces a different environment, and thus requires different capabilities in order to achieve the required fit to its environment. The complete firm, competing in the market is subject to a strong influence of the market forces, such as influences of the competition, the regulation in the market, imports, general economic situation etc., most of which would not bother the TSU, or have little influence.

The TSU is not directly influenced by the general economy and its ups and down – or at most as a secondary influence feeling the impact of lack or abundance of available funds for investment. To some extent, the seclusion of the TSU is also a protection from external changes, and the distraction they may offer. Once the recent investment round is completed the TSU can enter into seclusion and focus on the R&D required to bring it to the next round of investment in the best form it can.

This is not to say that it can afford to ignore the out-side world. Rather, that it can focus on the goals at hand, and find the right balance between the employees focus on the goals, and the manager's attention dedicated to identifying threats and opportunities.

The complete firm is a market player and needs to have the ability to sense and analyze changes and potential dangers at all times.

More specifically, it is influenced by the market forces, influencing its ability to sell and generate income. Further it is influenced by the regulation, determining its ability to operate, and influencing directly its cost of doing business. Another important aspect, this time a close aspect to that of the TSU, is the demand for growth by the shareholders. Last but not least is the influence of the financial economics as it would influence the complete firm on the revenue as well as the expanses sides of the ledger. It can influence the cost of work and materials, the cost of finance, as well as the money available for spending in the market and thus the income.

FUTURE RESEARCH DIRECTIONS

The research of the nature of the firm is vast and multidisciplinary. The topics of cultural and organizational aspects of firms are on-going and are in flux regarding research and practice. Since these topics have impact on economic growth/stability they are of international interest for academics and practitioners, and research shall continue regarding better understanding of efficient, environmentally fit, and manageable structure, as well as effective cultures and how to foster and preserve them. Issues such as the differences between Complete Firms and TSUs are not widely researched, but this may change as the interest increases in the transition from TSU to a Complete Firm.

Figure 2. The Complete Firm's environment

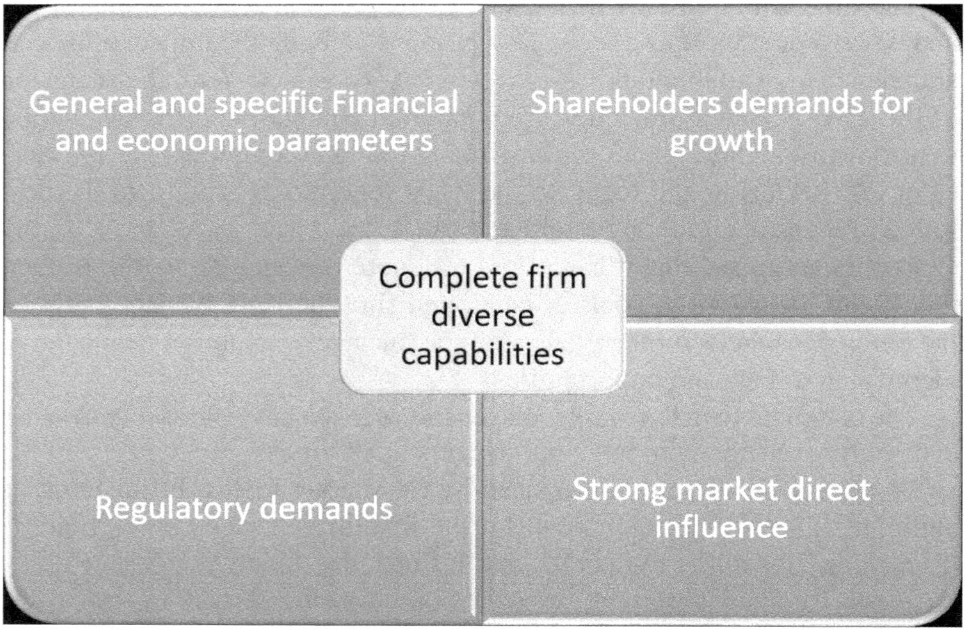

CONCLUSION

We have explained the concept of the "complete firm" as compared to that of the TSU. We have discussed the differences in the environment in which each type is operating and the resulting different demands. While we have seen that there are different ways to differentiate the TSU and the complete firm, we suggested a differentiated way to distinguish between the two (Santi and Neves, 2020).

These demands in turn influence the needs to "fit the environment" as it is expressed in the organizational and cultural structure of the different entities. The need is also expressed in the operation and the required managerial capabilities of each one. We suggested models to explain the needs of the firm, and to present the difficulties it faces in the transition.

For each there are advantages and dis-advantages as we discussed and depicted.

The difference in the required capabilities derived from extraneous forces and demands, between the TSU and the complete firm creates the gap – the topic for the next chapter.

REFERENCES

Audretsch, D., & Keilbach, M. (2004). Entrepreneurship Capital and Economic Performance. *Regional Studies*, *38*(8), 949–959. doi:10.1080/0034340042000280956

Barr, P. S., Stimpert, J. L., & Huff, A. S. (1992). Cognitive change, strategic action and organizational renewal. *Strategic Management Journal*, *13*(Special Issue), 15–36. doi:10.1002mj.4250131004

Baum, J. R., Locke, E. A., & Smith, K. G. (2001). A Multidimensional Model of Venture Growth. *Academy of Management Journal*, *44*(2), 292–303. doi:10.2307/3069456

Delmar, F., & Shane, S. (2006). Does Experience Matter? The Effect of Founding Team Experience on the Survival and Sales of Newly Founded Ventures. *Strategic Organization*, *4*(3), 215–247. doi:10.1177/1476127006066596

Gompers, P. A. (1995). Optimal Investment, Monitoring, and the Staging of Venture Capital. *The Journal of Finance*, *50*(5), 1461–1489. doi:10.1111/j.1540-6261.1995.tb05185.x

Hughes, M., & Morgan, R. E. (2007). Deconstructing the Relationship between Entrepreneurial Orientation and Business Performance at the Embryonic Stage of Firm Growth. *Industrial Marketing Management*, *36*(5), 651–661. doi:10.1016/j.indmarman.2006.04.003

Jaskyte, K. (2004). Transformational Leadership, Organizational Culture, and Innovativeness in Nonprofit Organizations. *Nonprofit Management & Leadership*, *15*(2), 153–168. doi:10.1002/nml.59

Korbet, R., & Analyst, M. (2019). *The State of the Israeli Ecosystem in 2018*. Start-Up Nation Central.

Nason, R. S., & Wiklund, J. (2018). An Assessment of Resource-Based Theorizing on Firm Growth and Suggestions for the Future. *Journal of Management*, *44*(1), 32–60. doi:10.1177/0149206315610635

Owens, T., & Fernandez, O. (2014). *The Lean Enterprise: How Corporations Can Innovate Like Start-Ups*. Wiley. doi:10.1002/9781118854006

Penrose, E. T. (1959). *The Theory of the Growth of the Firm*. John Wiley.

Porath, A. (2017). *Supporting Innovation Through National R&D Programs: Emerging Research and Opportunities*. IGI Global.

Santi, C., & Neves, G. T. (2020). Startups and prominent business. In A. G. Rodrigues (Ed.), *New and Future Developments in Microbial Biotechnology and Bioengineering* (pp. 209–223). Elsevier. doi:10.1016/B978-0-444-64301-8.00010-X

Sarma, S., & Marszalek, J. (2019). New Venture Growth: Role of Ecosystem Elements and Prior Experience. *Entrepreneurship Research Journal, 10*(1). . doi:10.1515/erj-2018-0215

Zahra, S., Jennings, S., & Kuratko, D. (1999). The Antecedents and Consequences of Firm-Level Entrepreneurship: The State of the Field. *Entrepreneurship Theory and Practice*, *24*(2), 45–66. doi:10.1177/104225879902400205

Zahra, S. A. (1995). Corporate Entrepreneurship and Financial Performance: The Case of Management Leveraged Buyouts. *Journal of Business Venturing*, *10*(3), 225–247. doi:10.1016/0883-9026(94)00024-O

KEY TERMS AND DEFINITIONS

Complete Firm: A company for profit, deriving its resources mainly (not exclusively) from its daily operation as revenues from the business it is conducting. Deriving the revenues from the business requires a complex collection of different activities and this requires in turn complex set of managerial capabilities.

Environmental Fit: By defining an enterprise or any organization, a definition of the environment in which the organization operates is formed. Borrowing from biological evolution, the level of adaptation to the environment of the organization is the environmental fit of the organization. The higher the fit, the more adapt the organization to the demands and opportunities of the environment in which it operates.

Organizational Culture: For the purpose of our discussion here. The organizational culture shall refer to the forms of the interactions between the enterprise organs and the collection of customs and values of the enterprise.

Organizational Structure: For the purposes of our discussion here the organizational structure of an enterprise, refers to the organs in the enterprise and their interactions.

ENDNOTES

1 For more details regarding the SME Instrument program please see: http://ec.europa.eu/research/participants/portal/desktop/en/opportunities/h2020/ftags/sme_instrument.html#c,topics=flags/s/SME/1/1&+callIdentifier/desc

2 In fact, as this is the main topic of the original management literature, no quotes are presented here.

Chapter 5
The Nature of the Gap

ABSTRACT

This chapter describes the gap between the complete firm and the TSU and the challenge it presents to the enterprise and to the entrepreneur/owner. First, the different aspects of the gap are presented. The description of the required changes follows. This includes the establishment and growth of missing units/activities in the TSU required to make it a complete firm along with the demands this change places on the entrepreneurs/owners. The issue of the required learning relating to the required accumulation of managerial capabilities is presented. This leads to a discussion of the potential crisis that may result from the need to address the new demands. Some potential risks for the entrepreneur are presented regarding the change.

DOI: 10.4018/978-1-7998-5685-6.ch005

INTRODUCTION

The fact that there are organizational, cultural and thus managerial differences between the TSU and the Complete Firm does not automatically mean that there are problems in the transformation from the first to the latter. In order to present the difficulties and risks involved in the transformation, the nature of the gap should first be explained.

It is important to recall that the TSU is in the process of creating value internally in order to further develop. This is part of the background to the transition, and it is important to understand where it can impact the entire activity.

Only once the nature of the gap is clear, can the demands on the entrepreneur / owner be presented and explained. The demands as will be explained in detail below, require the entrepreneur to not only adapt to a new way of doing things, which is difficult enough in some cases, but actually to develop additional managerial capabilities. That development of managerial capabilities is placed in most cases not on a single person, but rather on the management group as a body. However, it requires some learning process from all in order to participate in decision making related to the new disciplines introduced.

Thus, it is important to understand that the transition from one mode (TSU) to another (Complete firm) is more in the nature of a revolution than a calm continuous process. This can create a "breaking point" for entrepreneur and an exit point for some from the enterprise.

THE NATURE OF THE GAP

In the chapters above we described each type, the TSU and the Complete Firm. In this chapter we shall describe the gap between the TSU and the Complete firm, from all the aspects described above. The difficulties in covering the gap and the actions required to do so. As in most fields of management, here as well there are several possible solutions for the difficulty in overcoming the gap. We shall focus on the main solutions and the reasons for selection of each one, in the following chapter.

For clarity purposes we shall first define what is not the gap we are discussing here. There are different gaps in the value creation process of a TSU. There can be a gap between each of the development stages described below as TRLs. There can be a gap between a small entrepreneurial group working on an idea in their free time, and a case in which some people are being employed and paid for their work. Each of these gaps, can be crucial for the TSU to cross successfully, and TSU that fail to do that – perish, or become something else.

These gaps in the value creation, R&D process, are not the gap we refer to here. Our gap as mentioned above and further described below is not financial in nature. It is also not a question of size and additional tasks, but rather a radical shift in the concept of the firm and its culture.

It is important to remember that the issue of the gap, and the differences in the culture, as well as the organizational structure of the complete firm and the TSU, influence the atmosphere in which the employees and the entrepreneurs operate. The move from one to the other results, among other changes, also in changes to the organizational culture, and the work environment, and this is not always a welcome change to the entrepreneurs.

It is important to recall at this point, that the "perpetual / repetitive" entrepreneur is in fact a person who likes the special work environment of the TSU and thus prefers to move from one TSU to another, as the first one matures into a complete firm. One needs to take that into account, that while the process may not be easy to the employees who may also wish to move to another TSU and continue to work at a similar environment, it is more problematic to the entrepreneurs, who in many cases must lead the change. We can all recall that in the movie "The Intern" with Robert De-Niro and Anne Hataway, the manager of the successful SU, was pressured into selecting an experience manager to take over the firm that she had founded by her board, as they felt that the firm required a more experienced manager to face the challenges of its growth and expansion. While the firm in the movie was already a complete firm it was in the process of growing to a more mature level and the main challenge was making sure it could remain efficient and provide the required service to the clients.

There are many ways to describe the development process all the way from the early idea to the final product sold on the market. One of the more widely accepted is the Technology Readiness Level (TRL) of the EU[1].

The TRL gives each stage a definition regarding its progress towards the market. Please see below the quote from the EU document: "

- TRL 1 – basic principles observed
- TRL 2 – technology concept formulated
- TRL 3 – experimental proof of concept

- TRL 4 – technology validated in lab
- TRL 5 – technology validated in relevant environment (industrially relevant environment in the case of key enabling technologies)
- TRL 6 – technology demonstrated in relevant environment (industrially relevant environment in the case of key enabling technologies)
- TRL 7 – system prototype demonstration in operational environment
- TRL 8 – system complete and qualified
- TRL 9 – actual system proven in operational environment (competitive manufacturing in the case of key enabling technologies; or in space)"

While the above seems to fit many processes there are some points that need attention:

It fits a certain process that starts with fundamental research. The first two TRLs are not relevant when you are dealing with a development of a product, that started from a need in the market and not from basic research, that led to a technology etc.

Think for example of the development of a suitcase on wheels. There were no basic principles observed, or a technology concept formulation. Another problem is that of course the TRL system, does not fit the development process of pharmaceuticals or even some other related fields, where due to regulation the process and stages are different.

This is not to disparage the attempt to formulate a comprehensive description of the process. The fat that it is at least partly successful can be derived from the SME Instrument program of the EU Horizon 2020 program, which makes use of the TRL to describe the level of the projects accepted to its second phase, and the progress they need to make during the proposed project life[2]. This program is a relevant example, as part of its aim, is to promote the company making the proposals to the next phase of investment by maturing it along the development process.

For our purposes here, we can assume that the process (TRL), is correctly described or at least in a relevant manner. In that case we can see that the end is just where we describe our gap. We have finished the development, we are ready for manufacturing, may even have the manufacturing file in hand, now all that remains is to set up the manufacturing process, start ordering inventory, manufacture, sale, market, etc., this is not included.

Just the short list in the last paragraph above demonstrates the multi-disciplinary nature of the gap. From a limited marketing understanding, combined with heavy technological understanding, we need to move into something that encompasses marketing as a specialized and professional activity, same for finance, inventory and HR management, production management the logistics and other activities, that makes life very complex.

Moreover, the new disciplines do not always arrive separately, give you time to adjust to the new arrivals before other come up as well, they may all appear at the same time. Even if you think that you can out-source some of them, and you (think it and do it), you will still need to manage it. You need to find the sub-contractors, define the work, and protect your knowledge, contract them, check the quality of the work, and manage the interfaces of the contracted activity and the rest of the firm. This is not easy.

It is especially not easy, if this is the first time you do it, and if you have never tried the specific tasks before. There may be problems that you never encountered before and therefore have no ready solution for, certainly not when your sub-contractor must deal with them.

In the sections below, we shall try and divide the explanation regarding the gap into sub-topics, to make it easier to understand, and perhaps reduce the confusion. It is in many cases a shock. This is part of the reason many firms are afraid to try and cover the gap, and instead prefer to avoid the move entirely, as we shall discuss in the solutions section below.

An important aspect is of course that start-up, when ready to move to complete firm stage, is normally already used to fast growth. Most of the start-ups that succeed, experience a phase of rapid growth. So, they know how to grow and recruit, and in most cases how to train and assimilate the new members. The issue is that the move towards a complete company requires growing new activities, while the start-up growth was mostly focused on increasing the R&D effort.

Having generally described the challenge, let us move to a more detailed description of the gap and the different manifestations it presents.

The Organizational Change Required

In this section we shall discuss the organizational aspects of the move from the TSU to the mature firm. One important aspect to remember is that in most cases the move from TSU to a complete firm requires growth. Only in rare cases in which it happened, that the R&D activity was downsized, and the new HR required for the move just completed the HR quota. Normally, even if there was some downsizing in the R&D activity, it in no way compensated by the recruitment required for the move. In most cases there was a huge effort to maintain the R&D resources, that the firm invested so much in recruiting, and did not want to see them leave[3].

So, there is growth, but we need to better define this growth. A TSU grows most of the time, either continually or in leaps. However, the TSU growth is normally, more people of the same kind to perform adjacent tasks, R&D in different aspects, and therefore has little impact on the structure or organizational culture of the firm. That is not to disparage the growth phenomena of TSUs during their R&D period. A growth step from 5 people to 40 in a month is a huge step and requires a substantial effort. It is a major move and a significant management challenge. However, in many cases these are people dealing with more or less the same or closely and similar challenges, in the R&D area mostly and are thus defined mostly as "more of the same". It is the size rather than the complexity that constitute the challenge. It still requires re-definition of the organizational structure, management culture and reporting procedures.

In the gap, the growth challenge is at least two dimensional, as mentioned above, it is in some areas more of the same, but it is also an introduction of new tasks, new people, new structures etc., thus more complex. One should recall that it may change the organizational culture as we discuss below.

The complexity means that it is not a "linear" growth in which more departments are created, new managers are nominated, and some former heads of the departments are promoted to higher management level. Such a linear growth will have little impact on the organization culture, but this is not our case.

In the multi-disciplinary growth scenario, we discuss the addition of new disciplines in which another culture dominates the units created. The culture in manufacturing is very efficiency oriented, expenses, times, maintenance etc., are all closely watched. Work is in shifts, and all must be on time. It is not individualistic-like development work, in which in many cases each can work on its own part until the rest of the team is ready to interface. Even teamwork in that context is limited. You cannot start part the line while waiting for the rest of the production line workers to come. That dictates a new organizational culture in the manufacturing department that did not exist before in the TSU and that is suddenly in the firm.

It is not just the case of manufacturing, finance and legal are more formal than R&D and add a side of formality to the operation. The need to transfer information from one unit to another, in a way that takes into account that the recipient may be less proficient in the technological aspects of the information, and thus requires it to be simplified but still informative, changes the internal communication between the units and impacts the overall organizational culture.

Therefore, the addition of the new units with the new disciplines will impact the organizational structure. These units may also be structured with a different hierarchy to the older units, and the interfaces, when e.g., there are joint meetings, become complex. It is in most cases, impossible to create complete separation between the units, the old and the new and thus it has to be considered that the new additions will be known all over the firm and that the changes will be scrutinized and compared to the status and conditions of the older employees.

Changes in the demands from the workers, different communication protocols, added formality, confusing hierarchy, no wonder that certain people prefer to move to another TSU, where they feel more at home with the organizational culture, than stay in a place where they used to feel that way, but has changed.

Figure 1. Linear vs. two dimensional growth

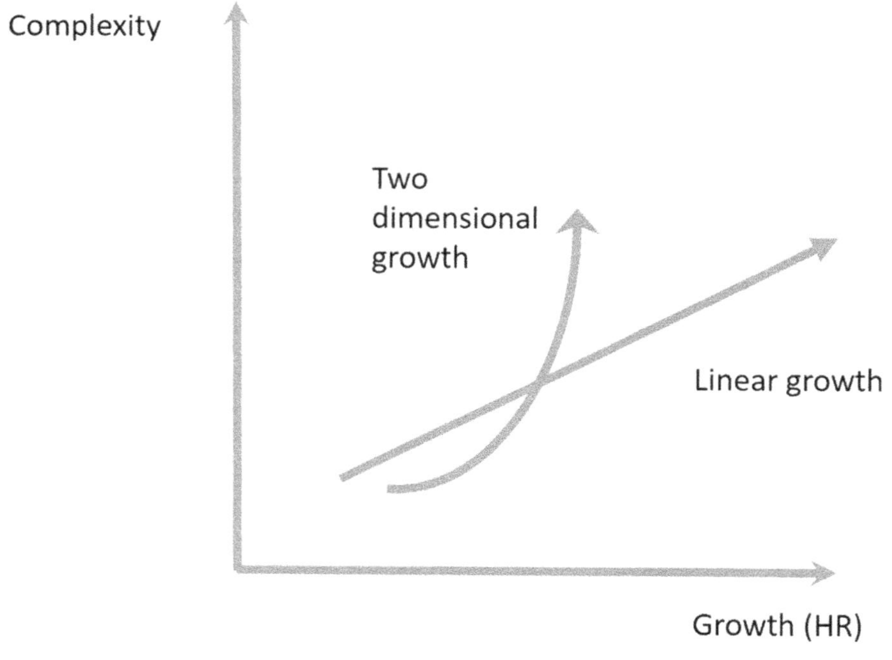

This is an additional challenge to the gap, to maintain enough of the people from the TSU so as to keep the intellectual property in the company. This is not a simple task, as the finance is often not the relevant means to do it.

We discussed the need to grow, and that growth is both two dimensional, and rapid, and thus has an impact on the organizational structure and culture.

The "Missing" Parts: Units, Operations, Capabilities, and Processes

The growth of the new units is more complex than can be imagined at first. The need to build new units with new responsibilities requires some understanding regarding the function of such units. It is very difficult to build a production unit, without first understanding the structures of such units, their needs regarding infrastructure and HR and the logistics related to them. Therefore, for the graduates of an R&D unit, the creation of a production unit is often considered as an impossible task (Denison, 1990; Brown and Eisenhardt, 1998; Baum, at all, 2001; Sarma and Marszalek, 2019).

One way is to copy the structure of similar units from other, preferably similar organizations. It can be done using the experience and contacts of the founders, via consultants, or support from the ecosystem. Another way is to recruit an experienced manager in the field (e.g., experienced production manager) and let the experienced manager form the unit, create the inside regulations and procedures as per the manager's experience.

While the first alternative suffers from all the possible troubles of a task performed by inexperienced personnel, it does however, allow the founders some control over the process, and lets them transfer the control when they are ready to the manager of the unit.

The second option allows for better performance, as the formation is performed by an experienced manager, however, it means that the founders lose control over the process at the outset and have to be able to delegate the entire operation to someone else.

Both options are not fool proof. The founders may make mistakes in the first option in the formation, the regulations, and procedures, in the HR and other infrastructure. In the second option the most common mistake of the founders is to intervene in the formation process and hinder the expert manager in forming the unit.

The second option requires more maturity on the side of the founders, and probably would be preferred by the experienced entrepreneurs.

There is however difference between the different units. Production and maintenance would probably be closer topically to R&D than sales and marketing, or finance. There could be therefore, a difference in the formation of the different units. The founders may elect to hire an expert manager for sales and marketing, and for finance, while starting with sub-contracting for the manufacture and slowly integrate the production into the company, under their own control, until ready to recruit a manager and transfer the control.

The Different Requirements From the Owners/Entrepreneurs

One of the major challenges in a TSU is of course the management. It is like frying an egg while riding a galloping horse. The work environment is changing very rapidly, more people enter the firm, and more departments appear. However, as mentioned above most of the TSU growth is in one mode - the R&D. The challenges that the manager faces, are to get accustomed to a changing environment that includes growing a management hierarchy. When leading a small group of 4-5 people, the manager can directly manage them, oversee every aspect of the work, and go down into the finer details. This in many cases is the biggest fun for the entrepreneurs, especially those that do it for the first time. Most of the first timers who come from the industry, used to be managers of R&D teams and are more used to this aspect. They know how to deal with the problems and how to manage the team (Zahra, 1995; Cohen, 2006; Delmar, and Shane, 2006).

When the firm grows the entrepreneurs, especially the technical ones are pushed up, as more management layers are formed, normally below them. It is easy to understand that you cannot manage and task a team of 30 or 50 employees the same way you do for 4 or 5.

Now, after the growth, the manager has to cope with managing without direct involvement in all aspects. At most, the management is done in a department or unit level, setting the unit tasks and letting the unit managers divide the tasks between their team members. The hands-on management style is no longer relevant, and a more remote style is required. However, this still relates to the technical aspects, and thus the manager's area of expertise.

The remote style requires to be able to manage by reports and plans and to relay and trust the other managers to do it just as good as the original entrepreneur. This is part of the challenge of the growth at this time. However, in the two-dimensional growth the manager has to delegate responsibility in areas he is less familiar with, and thus has less confidence in the decisions taken. This is a higher jump for a leap of faith for the founders.

There are two challenges involved, one relates to the managerial capabilities, the ability to manage by reports; the second is a more psychological aspect, trust, and also relates to the ability to keep of actions and sections that are being managed by others.

The challenge of moving from the hands-on management style to the remote style relates more to the technical capabilities of the managers, but also has an emotional aspect of course. The second, regarding keeping of the actions and units managed by others, is less technical and almost purely emotional.

Figure 2. Nature of the Gap

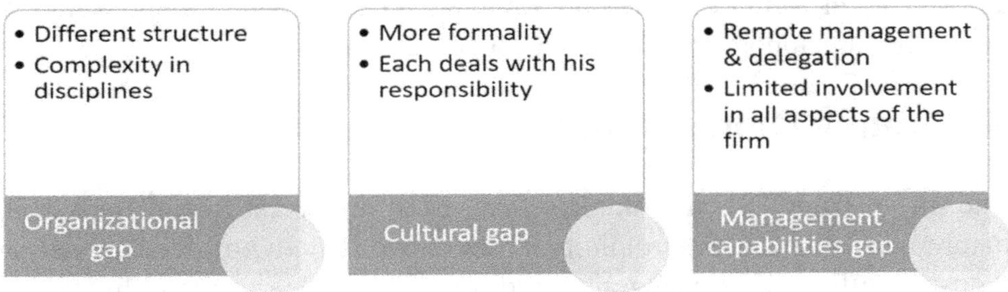

The Resulting Conflict: Potential Crisis

The organizational change required in order to make the move to a complete firm, brings a conflict both functional and emotional to the firm, and especially to the management.

On the one hand, this is what they, especially the founders have worked for all these years. This is the dream they have sold to the investors and pitched to countless others.

On the other hand, it is the end of their way of life up to this point. Things will never be the same, they will never go back. The only resembling environment they can ever operate in, is another TSU.

We need to understand some of the psychology involved in the move. The founders have started their business, in order to create an entity that will bring their invention to the market, to create a new thing, to create a living business. It is less important why they originally intended to create the TSU, but once it was formed, and for a certain period of one to several years, they have in a very intensive way, taken care of it. They have invested emotionally as well as otherwise in developing the business until the time for the change.

At the point of change, the reasons for starting the TSU are less important than the emotional investment. It is therefore a very conflicting moment when they have to decide how to let go. It has been compared to letting your children leave home. On the one hand' one is proud and happy and on the other hand, it is the end of an era, and there is some sadness.

At this point for some it is easier to leave, and not face the daily change in their limited involvement, in the struggle to re-define themselves and their role in what used to be very familiar environment.

As mentioned above some prefer to leave, while others struggle to make sure the change is handled correctly and that their cherished charge is protected. This is also evidenced in the choice how to make the move. The M&A with a larger organization in which the founders have no place except as part of the technical support, maybe easier for some.

Sic Transit Gloria Mundi[4]

The transit from TSU to complete firm, is an important factor. Both in its nature and its speed. The rapidity of the move is both enforced as well as a desired effect. This change is rapid many times due to the obligation to complete the transformation and inject income into the firm before the most recent investment is spent.

The transit is therefore required to assure the continuation of the activity of the firm, but it cannot be done rashly, as the change will affect the ability of the firm to survive in the middle and long terms.

The transit therefore requires that the company move from the organizational structure of the start-up to the organizational structure of the complete firm. However, there is a middle ground which is the most difficult stage of the management. In many cases the firm can and does begin to create some revenues from its future normal operations. These cases take the form of paid beta-site, market probing, trial sales and similar activities in which the firm engages.

The purpose of these activities is in many cases a further proof of concept, this time at market level, intended to show the investors, and many times also the founders, who have doubts in spite of having sold the idea originally, that there is a good potential for a market uptake.

This is an important step for the firm, for, while it does not generate in most cases, enough income to sustain the firm, it does indicate the future. The indication signals the future additional needs the firm will have to respond to, in order to be in business – guidance, customer service, documentation updates etc., while also indicating to the management the correctness of their expectations. This is done mostly regarding the market response as well as the readiness of the product to enter the market. In that transient mode, however, the structure of the firm is not really changed. The additional tasks and updates are normally left to the R&D department, where the required knowledge to design the guidance, solve the customer support issues and update the documentation lies. In some cases, the R&D department will learn how to instruct the internal sales department based on their direct contact with the clients during the beta-site.

These additional functions, therefore, are performed in the start-up way, by expert, in the technical domain, but with little specialization in the specific task, except perhaps in the documentation case. That is to say that customer support may not be as emphatic and polite as a professional one but will most certainly be able to solve all technical issues, and the guidance may lack in pedagogic knowledge, but not in the technical expertise required.

There is no organizational change and the specific persons delegated to perform these tasks are expected to revert back to their duties asap. However, they give an indication for the future changes that will be required.

This stage may be further extended in time, if a long-term client is acquired, demanding on-going service and therefore demanding also in time more professionalism in the support services. This may lead to a staggered move towards the complete firm, which is performed in stages, hence the staggering, and mostly follows the demands of the client/clients.

FUTURE RESEARCH DIRECTIONS

Potential future research regarding this chapter would focus on the learning demands on entrepreneurs and managers in general, as well as on the different phases in a firm's life (Jiménez, and Sanz-Valle,2011; Porath, 2017).

Another aspect which is well researched regarding value creation in TSUs specifically and in firms generally. This will continue to be of interest as value creation is TSUs is the key for further investment and thus the continuation of the TSU.

CONCLUSION

We opened the chapter with a general sketching of the nature of the gap. We than moved to a more detailed description of the nature of the gap, explaining the difference in the growth mode between the linear growth of the TSU, and the two-dimensional growth in the change to a complete company. We mentioned the resulting implications as well.

We referred to the change in the character and operation of the different types as a GAP, as this is both a clear divide between the types, as well as being a potential breaking point, from which the firm may not emerge whole. It is also a gap that needs to be filled – funds are required for one. But on that, please see the chapters below.

We mentioned the pioneering role, entrusted in many cases to the R&D department, as the source of knowledge, as well as the only source of available HR, before the other department with specialized personnel are established.

We presented the three aspects of the gap, organizational, cultural and management capabilities. Following that, we outlined the managerial impact the changes have, and deduced from it, part of the managerial conflict the founders meet when the change occurs.

In the next chapter we shall present potential ways to overcome the problems in more detail.

REFERENCES

Baum, J. R., Locke, E. A., & Smith, K. G. (2001). A Multidimensional Model of Venture Growth. *Academy of Management Journal*, *44*(2), 292–303. doi:10.2307/3069456

Brown, S. I., & Eisenhardt, K. M. (1998). *Competing on the Edge*. Harvard Business School Press.

Cohen, B. (2006). Sustainable Valley Entrepreneurial Ecosystems. *Business Strategy and the Environment*, *15*(1), 1–14. doi:10.1002/bse.428

Delmar, F., & Shane, S. (2006). Does Experience Matter? The Effect of Founding Team Experience on the Survival and Sales of Newly Founded Ventures. *Strategic Organization*, *4*(3), 215–247. doi:10.1177/1476127006066596

Denison, D. R. (1990). Corporate Culture and Organizational Effectiveness. New York: John Wiley & Sons.

Jiménez, D. J., & Sanz-Valle, R. (2011). Innovation, Organizational Learning, and Performance. *Journal of Business Research*, *64*(4), 408–417. doi:10.1016/j.jbusres.2010.09.010

Porath, A. (2017). *Supporting Innovation Through National R&D Programs: Emerging Research and Opportunities*. IGI Global.

Sarma, S., & Marszalek, J. (2019). New Venture Growth: Role of Ecosystem Elements and Prior Experience. *Entrepreneurship Research Journal, 10*(1). . doi:10.1515/erj-2018-0215

Zahra, S. A. (1995). Corporate Entrepreneurship and Financial Performance: The Case of Management Leveraged Buyouts. *Journal of Business Venturing*, *10*(3), 225–247. doi:10.1016/0883-9026(94)00024-O

KEY TERMS AND DEFINITIONS

Perpetual/Repetitive Entrepreneur: An entrepreneur who like the entrepreneurial stage so much that each time the TSU matures, the entrepreneur leaves the enterprise in favor of starting another one.

Value Creation Process: For the purpose of our discussion this is the process in which by advancing along the steps of the development chain from basic technology to finished product in the market, a TSU increases its value for investment.

ENDNOTES

1 Downloaded from https://ec.europa.eu/research/participants/data/ref/h2020/wp/2014_2015/annexes/h2020-wp1415-annex-g-trl_en.pdf on 03.05.2019.

2 For more details about the program see https://ec.europa.eu/programmes/horizon2020/en/area/smes

3 However, there may be such cases. Even if there was a compensating downsizing effort, this is rather the exception to the rule. For our discussion here, we assume there is a net growth.

4 Thus passes the glory of the world.

Chapter 6
Overcoming the Gap

ABSTRACT

In overcoming the gap, there are several strategies that can be considered. The two main strategies discussed in this chapter are the merger into a complete company (M&A) and the development of the missing parts with the required managerial capabilities. The M&A is presented and explained with the advantages and risks involved in taking this option. The M&A is also presented as a potential avoidance strategy, which circumnavigates the problem by merging the TSU into a complete company that has all the required units and the required capabilities to manage the complete firm. The second strategy is discussed with its own challenges in creating the new units, learning, and developing the managerial capabilities required. The opportunities advantages and disadvantages of this strategy are also presented.

DOI: 10.4018/978-1-7998-5685-6.ch006

INTRODUCTION

In the previous chapters we have discussed the nature of the gap, explained where it comes from, and the differences between the two entities. In this chapter we shall discuss the different means of overcoming the gap. As can be expected there are different possible solutions, and it is impossible to cover in one book all the variations on the solutions.

In order to overcome the different variations and challenges, we shall discuss the two main variations, to use existing infrastructure via a merger to a complete company, ready to offer the use of its facilities; or as alternative, to develop the different capabilities by the TSU as part of the change. There are middle cases, which shall be referred to, below.

Different Possibilities to Overcome the Gap

As mentioned above there are two classic possibilities: The TSU can elect to merge into a complete company (normally a larger corporation) and use the facilities the complete firm has to offer. In the sections below we shall discuss the pros and cons of that specific solution, as well as the cost and the rationale behind it. We shall briefly mention the process itself.

Another solution is the self-development of capabilities, infrastructure, activities, and actions in the TSU. It is in fact a family of solutions that is agglomerated under this specific heading. This alternative includes the development of some of the capabilities, or all of the new ones not internally, but as an out-source solution. The different levels of internal development are all grouped under this heading, and while we shall discuss the pros and cons of the internal versus the external development, we shall do so in a general manner, dealing mostly with the impact on the company.

In the discussion in this chapter when we shall refer to the company, we refer to the TSU, making the transition from TSU to a complete company.

Avoidance: M&A Into a Larger Firm

The process of merging into a larger firm, at least into a complete firm seems to be at first glance the simplest solution. In such a case, you instantly get access to the different departments, personnel, and managerial capabilities of a complete firm, without the need to develop them. If, as is usually the case, the complete firm is the larger entity, the merge will increase the volume of its activities, but would probably nor require it to increase the infrastructure significantly at first.

The merge also removes the anxiety of the TSU manager, regarding the need to make the move, recruit the resources and manage both the change, and the complete firm at the end. In the sections below we analyze the advantages and dis-advantages of such a move.

Process

The M&A process is comprised of incorporating the activities and to a less extent the HR of the TSU into the complete company. In many cases it will have two phases, the first will include the employment of the HR of the TSU inside the complete firm; and in the second phase some or all of the original employees of the TSU will leave the merged company. Their roles and knowledge having been passed to the workers of the complete firm – original or new. As in all mergers, the idea is that the new entity performs all that was performed before by the original separate entities but with less resources and more efficiently.

The first phase may not even include all workers of the TSU, focusing on those that have the key knowledge to the TSU's innovation. This is normally part of the contractual obligations of the TSU founders stated in the M&A agreement.

Advantages

As mentioned above there are some advantages for merging into a complete firm with the ability to absorb the TSU. Some of the major advantages are:

1. Gain immediate access to the seasoned experienced and active complete company infrastructure (time);
2. Avoid the need to recruit the resources and create the infrastructure (effort);
3. Avoid the need to create the management structure and train it in the business (effort);
4. Gain immediate market access, start bringing in business (finance);
5. Avoid the managerial headache in managing the system (effort).

The main advantage in many cases especially for entrepreneurs who have never done the move before, and who are looking to start their next TSU, is the last point. For people who are not interested in building and managing the future business for many years, but who rather like the TSU culture, the move is more of a must and less an adventure. This solution, which solves most of the problems, is fast and allows to move to more likable activities, is the ideal solution.

There are also advantages for the complete firm. In such a merge the complete firm receives an injection of innovation that would have been difficult in most cases for it to develop internally. It also removes in many cases a dangerous and innovative potential competition and gains an important strategic advantage over the competition in the market.

Disadvantages

On the other hand, there are also some dis-advantages to that move. The merge removes most of the special character and the characteristics of the original TSU, including those that were not related to the TSU stage of the firm. Some of the main dis-advantages are:

1. Damage to the TSU brand that is swallowed up in the complete company, especially if the complete company is much larger than the TSU;
2. The danger of losing many of the key persons of the TSU, and with them a lot of the knowhow related to the technology;
3. Losing the control over the company, as the management at best becomes a combination of the management of both companies, and in most cases, especially if the complete firm is very large, the large company management will take over;
4. In cases where the entrepreneurs were interested originally in building a company which they could run and manage, that dream is exchanged in many cases for cash / shares.

For the complete firm involved in the merge there are also dis-advantages. The new entity to be incorporated may cause disturbances in the status-quo achieved over time in the complete company. The newcomers bring with them some organization culture remains that may interfere with the organization culture of the complete firm. This step could cause some internal problems also, as some management roles need to be found for the managers of the

TSU. These new positions may come at the expanse of some of the managers in the complete firm, and cause problems, HR changes etc.

The Mature Firm needs to be able to know how to manage the move in order to benefit the most out of it and avoid the pitfalls. As the larger of the two, and the one absorbing the TSU, it has the responsibility for the move.

Risks

Not all M&A are successful. In the cases of larger firms (automotive cases are better known), failure can lead to a division of the firm back to the original firms that merged. However, in the case of a TSU such a retreat could be fatal for the firm, as the TSU is not self-sufficient (Scalera, 2014; Folz, 2003).

Failure in our case, can mean that most of the original TSU and at least some of the founders will leave the united firm. If they are prevented from doing so for a period by the M&A agreement, they will do so at the end of the freeze period, but if the M&A is not successful, they will leave. It could also have a negative impact on the complete firm, in influencing the organizational culture, and perhaps creating disturbances in the complete firm. These disturbances could cause economic and organizational damages.

The results of the M&A are process dependent and require at least on the part of the complete firm, assimilating the TSU, experience and understanding of the process in order to manage it. This is definitely a risky move, but some firms, have developed it into a source for growth and innovation (e.g., IBM).

Developing the Missing Parts

The main alternative to the M&A is the self-development of the required capabilities. That would include the need to recruit the resources required for the move, as well as the ability to manage it. In this section we shall discuss the challenges the TSU faces and especially the entrepreneurs and managers in trying to effect the change themselves.

Challenges: Development of New Capabilities; Replacing the Leadership; Developing New Processes; New Assets; Shifting the Focus

The key to the move is the managerial ability of the leading team in the TSU to manage the move. The ability to manage the change is a prelude to the management need to manage the complete firm after the change.

As any managerial capability it consists of the ability to raise the require resources, the financial and HR to implement the change; the ability to chart the change and lead the company through the changes controlling the different moves.

There is an element of self-development in the management of the change and later the complete company, however, it is a great challenge.

Often, the change requires growth in the management, additional capabilities, and additional responsibilities, require an enlargement, more directors, more managers, and a wider management basis are required. In some cases, wherever it is possible, a preferable way to implement the change is by recruiting the managers of the future activities and units, and to let them, together with HR to recruit their workers. This way at least the personal element of the managers and their employees is taken care off.

Such a move may have the advantage that the new workers, being recruited by their managers, would have no authority issues, or less than could be expected, but this requires experienced managers. Who knows how to recruit their workers and how to set the work processes and manage them?

Thus, recruiting the mangers solves some of the problems, but requires experienced managers who can set up operations, and manage them. These people are more expansive than less experienced managers. They require a lot of experience, are probably recruited from the outside and not promoted from within – this could cause some problems with the older workers, seeking promotion.

The recruitment of the managers allows the entrepreneurs some relaxation, as they do not need to set the operation themselves, or recruit people for positions, they are less familiar with, and may have trouble defining the requirements for these positions. They are also exempt from setting the process and operations of the units. However, they are not exempt from raising the funds required for the move. Not to forget, their own difficulties in delegating.

Opportunities

One opportunity is the ability to mature the firm into the economic entity the entrepreneurs dreamt off when they set out to form the TSU, while maintain control at least to some extent of the firm. The entrepreneurs are still required to raise the funds for the move and to make sure that the resources required are available.

Thus, for entrepreneurs who wanted to form the TSU in order to lead the economic entity, and to create a new enterprise, it is a wonderful opportunity. If done in a way compatible with the managerial capabilities of the entrepreneurs, leaving the new activity and disciplines open to newcomers, it can be a positive experience.

The TSU veteran managers are however, required to relinquish some of the hands-on management, and involvement, they used to have. They will need to delegate the areas that are not under their control to the new managers, and to be able to trust the newcomers in the domains where the entrepreneurs understanding is the least. They may need to delegate over all control to newcomers, who become their new bosses.

This is also an opportunity for the entrepreneurs to see "how it is done" by experienced managers. If they will not stay in the new complete firm, but opt to start a new TSU, they will know how to do that move, or to be better prepared for it in the future.

Risks

This alternative is also process dependent. The need to make the change is derived by financial and strategic constraints. There is a need to enter the market and establish a complete firm that is self-sufficient. If the process fails for any reason: lack of resources, delays that prevent timely market entry, HR tumult leading to a significant of key workers leaving etc., this could lead to problems and in rare occasions event to a closure of the TSU.

The fund raising may not be sufficient for the required move or may take too long. This will delay the market entry, reduce the ability of the firm to generate income and thus cause financial problems, that may lead to dissolution of the TSU. Or they may allow the change, just not enough to reach break-even point or even financial independence from investment.

The move may be planned in such a way where the entrepreneurs select the workers or promote from their own R&D personnel and the new workers do not facilitate the move fast or accurately enough, leading to similar problems as mentioned above.

There are also issues of the entrepreneurs, not delegating enough to the new managers, stemming the development of the firm, and thus leading to future problems in the process or in income generation.

Advantages

There are several advantages to taking this way:

1. The entrepreneurs retain more control of the new company, and as much as they can of their old roles. They may lose more of their share in the company when raising the required funds for the move, but they will retain at least some of the managerial rights, at least in the R&D and technology related activities;
2. The entrepreneurs can better control the change, the rate of the change, and the impact on the workers that have been with them since the early days;
3. The gradual change may also help them get used to the new organization, the new modes of operation, and adapt to the new organizational culture;
4. The above point will also thus reduce the rate of change over in the HR.

Disadvantages

Naturally there are some dis-advantages to developing alone to a complete firm:

1. The process requires management, and if the team has no idea how to do it, it may go wrong. People may leave, the company may lose efficiency and may fail entering the market with disastrous results;
2. The process may take longer than the alternative mentioned above, with similar bad results;
3. The process requires some delegation of management to new managers with whom the entrepreneurs are naturally less familiar in general, than with the existing members of the firm;

4. While in an M&A, the reimbursement on the move is evident and normally part of the M&A agreement, in the development process, there is funds raising in which the entrepreneurs pay with part of the shares in the TSU, while the increase in the firm value, is contingent on the success of the change, and the future bankability of the complete firm. There is no monetization normally in this option.

This requires some reference to the macro-economic aspects of each of the options mentioned for the transformation into a complete firm. The M&A has benefit for the TSU in the potentially rapid and smooth move into a complete firm, allowing simpler market introduction. At the same time, it allows the bigger firms to inject the innovation they sometimes find difficult to develop internally. Thus, it allows for benefits for both partners. However, in the macroeconomic point of view it will be, probably due to its efficiency, less creative in the HR aspects. Less HR resources, in the non-technological support professions – administrative, accounting, finance, logistics etc., would be required, as this would be at first based on existing capacities of the complete firm in the M&A, and later be only marginal additions if the capacity so requires. Further, it will reduce the number of small to medium size companies, as the process supports the increase in larger corporations, rather than small independent firms. This by itself, as is known, can reduce the robustness of the economy in the relevant region (Storey, 1991; Samila and Sorenson, 2011; Taleb, 2012).

The evolvement of the TSU into a full company would have positive outcomes for the founders and to the original investors, but it would also have more positive results in the macroeconomic point of view. The maturation into a full company would entail the creation of some HR infrastructure that is fixed, and this would in the overall HR aspect create more jobs than an M&A would. At the same time as it tends to create small to medium size companies, it also increases the robustness of the economy (Storey, 1991; Taleb, 2012).

FUTURE RESEARCH DIRECTIONS

The field of M&A in research is an interesting one. The importance of successful M&A can be understood when considering the number of such event occurring every year, and the economic impact (positive) when they are successful, versus the impact (negative) when they fail. Therefore, understanding the parameters of a successful M&A and how to prepare for it, and then execute it, are of interest both academically and for practitioners in business and government.

The field of the growth of the firm and the relevant means and different stages is one of great interest, and it can be expected that questions regarding the best ways to achieve it would be the focus of research in the future, as they already are.

CONCLUSION

There is more than one way to bridge the gap in the move from the TSU to the complete firm. The two main alternatives presented in this chapter are very distinct, but one should bear in mind, that there are also mid-way possibilities. In the case study presented in chapter eight, we shall see an alternative to the process that is a version of a mid-way.

The two alternatives presented in this chapter, are the M&A with a complete firm in which the TSU is spared the need to develop the organization and managerial capabilities of a complete firm and enjoys that of the merged into firm; and the process of developing the capabilities, the infrastructure, and resources of a complete firm.

The advantages and dis-advantages of each one, were discussed including the "price" the entrepreneurs would need to pay in each case. In the case of the M&A it is also important to remember that there is another entity involved and that the process should be beneficial for it, just as well.

The success of both options is dependent also on the way the new activities and units are developed, or the way in which the merger is affected. It is important to realize that this stage is a crucial stage in the life of the TSU. It is a determining stage and could lead either to financial success or ruin.

Combinations or mid-way alternatives are also available. It is important to remember that we are dealing with generalities and while most entities would adhere to the descriptions mentioned above, there may be special cases – outstanding in description and results[1].

REFERENCES

Folz, J. M. (2003). Managing two brands for success: Peugeot and Citroën. In B. Gottschalk & R. G. Kalmbach (Eds.), *Markenmanagement in der Automobilindustrie*. Gabler Verlag. doi:10.1007/978-3-322-92081-2_16

Samila, S., & Sorenson, O. (2011). Venture capital, entrepreneurship, and economic growth. *The Review of Economics and Statistics*, *93*(1), 338–349. doi:10.1162/REST_a_00066

Scalera, F. A. (2014). Internationalisation Strategies in the Automotive Sector. The PSA Peugeot Citroen and Fiat Auto Groups: A Comparison between the two Models. *JIEB*, (2), 28–37.

Storey, D. J. (1991). The Birth of New Firms – Does Unemployment Matter? A Review of the Evidence. *Small Business Economics*, *3*(3), 167–178. doi:10.1007/BF00400022

Taleb, N. N. (2012). Antifragile. Things that gain from disorder. Random House.

KEY TERMS AND DEFINITIONS

Firm Growth: The increase in the assets (tangible and non-tangible) of the firm. It can be expressed in several ways, increase in income, sales, profits, market share etc., or in the number of employees investment in infrastructure (sites, buildings, and machines), as well as in the intellectual property and reputation.

Merger and Acquisition (M&A): The incorporation of one enterprise into another, through ownership but mainly through joint management. While in most cases it would be the bigger company that acquires the TSU, for our discussion it is not the size that is important, but rather that the Complete Firm takes managerial control over the TSU.

ENDNOTE

1 As a professor in a college mostly for more mature students with more extensive life experience than 18 years old high-school graduate, I used to have to deal with such special cases. Many proved not to be special cases, but rather misinformation, but rarely there were exceptions.

Chapter 7
Government Tools and Assistive Tools

ABSTRACT

The understanding of the difficulties in overcoming the gap, with the importance of maturing TSUs into complete firms, makes the involvement of the government in assisting the transition important. General aspects of the conditions and restraints on government support are presented followed by the reasons for government involvement. The main need for government support, to overcome market failure, is presented and explained as is the importance of the TSU economically and socially. The requirements in terms of learning and developing capabilities as well as the difference between the needs of internal actions and out-sourcing is clarified. The difference between financial and non-financial support is explained, as are the differences between general support (e.g., standardization) and firm specific support. Some of the specific tools (e.g., beta-site support, marketing entry support tools) are presented. Their types and purposes are described in general, with a reference to other sources for those interested in reading more.

DOI: 10.4018/978-1-7998-5685-6.ch007

INTRODUCTION

The importance of the TSU for the economy as an engine for growth or at least stability in the economic system is fairly clear. It is therefore important for those in charge of developing the economy that the TSUs have a smooth transition to the next stage or risk the collapse of the entire sector of TSU due to lack of continuance. The special conditions of the TSU have found expression in the development of special tools to assist the TSU. There are private tools in which the market is filling the gaps in order to enjoy the economic benefits resulting from the TSU[1]. However, there are also cases where the free or private market fails to get involved. Due to high risks, or other reasons. In these cases, there is ground for government or public intervention to compensate for the market failure and assist the TSUs in the transition. That is the rationale behind the non-private tools. In this chapter, after discussing the reasons for public support tools and their characteristics as well as risks, some examples are given of such tools, not all of them are financial aid. It is also important to distinguish between the specific aid (financial or not) given to a specific firm, and general aid to support TSUs as a sector, or type of enterprise.

General Aspects of Government Support

When discussing government support, one has to keep in mind that there are several basic assumptions driving Government support activities.

Assumptions:

- Government support is based on public funding, and thus the same care and impartiality have to be used when employing measures using public funding;
- Government support should be aimed at the general public good and not favor specific individuals;
- Government support should be based on transparent procedures with pre-published criteria for award of support decision making;

Therefore, government support requires the creation of publicly known criteria for awarding the support, transparent procedures, and financial control over the utilization of the support. Further, as the special care required for public funding and the general good, would require that the support be given where there is a market failure. Otherwise, if the measures are employed where there is no "market failure", the government is actively competing with private players in the market, which is normally not in the public good.

The government support should therefore be transparent, have clear cut, published criteria, and aim to overcome market failures[2].

Therefore, the process of creating such government support tools, would require at first to identify market failures, define the measures to overcome them, and the criteria for awarding this support. Only then can the activity proceed to securing the finances.

There are some classical market failures, or gaps that require some government assistance. In the next sections some examples are given.

Market Failure Requiring Government Support

Creating an innovation ecosystem, where such an ecosystem did not exist – this challenge, the formation of an ecosystem requires government support. The market failure is evident as otherwise the ecosystem would be in place.

Another topic is the encouragement of innovation. This may have several aspects such as R&D support, support of entrepreneurial activities, support for the creation of a VC sector and others. This is a classical area for government support, as can be seen in the activities of the NSF and NIH, encouraging innovation through the development of knowledge and mainly R&D. As Lukkonen (1998, 2000) and Porath (2010) have shown, the encouragement of R&D and even generic R&D is important for increasing the wellbeing of firms, and the role of the government is to encourage firms to take Innovations risks, that the firms would not have taken without that support.

Another market failure is the entry into new markets. This type of market failure is derived from the lack of presence in the market on behalf of the firm. It exists for all firms; however, large firms have the economic tools to overcome the barriers and enter new markets. Small and medium size companies have a more difficult time in entering new markets. Government support in this case, identifies different types of new markets, local new markets, where government support is more cautious to prevent endangering the other players in the market; and foreign new markets, where the government sees macro-economic benefits in assisting the local firms in entering.

As mentioned above, the R&D support for the industry is easy to understand following large research projects, leading to large multi-national programs and strategies – such as the EU "knowledge-based society" (EU 2000). The support for existing selling companies, employing workers and supporting the economy is understandable (EU, 2003). But is it just as clear regarding the support for the TSU, or even the will to form more of them?

In the sections below we shall deal with the different forms of government support that can be given to overcome market failure and analyze which is most fitting for each case. But first we need to make sure that the support for the TSU and the move is in the public interest.

Understanding the Economic and Socio-Economic Benefit of the TSU

There is an economic benefit attributed to the TSU, which is comprised of several parts. The overall benefit derived from the TSU, may be bigger than sum of these parts.

The first part of the benefit is derived from the TSU being a small company, and the benefit to the growth and robustness of the small companies to the economy has long been demonstrated (Storey, 1991; Samila and Sorenson, 2011; Taleb, 2012). The importance of the TSU however, even in that capacity is bigger than the average small company. The TSU is more agile, and more innovative; therefore, it contains a higher growth potential than other new small companies. These companies are entering markets, some of which are already full of competitors, and these companies do not always have a competitive advantage, over the other companies already active in the market. The TSU is thus in general beneficial to the economic development and robustness.

The TSU is also in most cases, aimed, being an innovative technology company, at international markets. It thus has the potential to contribute not only to the economy in the capacity of a small and agile company, but also to the economic growth of the export, potential other types of SUs have to demonstrate under more difficult conditions. That potential, for an export often also expresses itself early in the TSU life, by attracting investment from abroad, increasing the inflow of foreign currency and the thus the economic potential growth.

The easy formation of TSUs, is a source of stability in the HR market. The STU tend to fail and disappear from time to time. However, if the formation of new one is easy enough, and there is a supporting ecosystem, the people dismissed from one find new employment in another one rapidly. Thus, the formation of the new ventures allows higher levels of employment, especially for the people trained in the right technical professions, but also for supporting professions.

Developing the Capabilities

The capabilities the TSU needs to develop are therefore complex and lay in two main domains, the practical domain and the managerial domain.

The Practical Domain

The practical domain refers here to the execution of the new activities that the TSU has never done before. One of the most obvious in most cases is production. TSUs having finished the development phase, at TRL9 for example need to move towards production. That in our discussion of the practical domain would refer to the actual actions of making the products, preparing, assembling, packaging and making them ready to be shipped to the market.

This new activity would require some investment, in the infrastructure, the recruitment of the relevant HR for the production, the setting up of the equipment and the training of the HR, including the mandatory quality control systems, and the start of the manufacturing process.

This can be done expansively by doing the entire activity internally, using the right experts to set it up correctly, arrange the transfer of materials internally to be economical and efficient, etc.

It can also be done in a cheaper way, by outsourcing the manufacturing to sub-contractors. This may be cheaper in the short term, but it has its own price. There would be knowledge leakage to the sub-contractors, it would require quality control to control the sub-contractors, and the over-sight to make sure that knowledge does not leak to the competition, or even some whole products. There are also long-term dangers, the loss of knowhow regarding production problems that the sub-contractor may need to solve, but will not share with the firm, alteration and usage of inferior materials or components etc.

However, the out-sourcing solution is cheaper as it does not require a heavy capital investment before the company can be sure that it shall have the income to cover that investment. It also seems less daunting to have that happen as it removes the need to actually manage the activity, but that is discussed further below.

For some companies, the out-sourcing is also a way of testing alternative production systems or processes, as one sub-contractor may use one process and another an alternative one. Comparing the production information from both would allow the former TSU to decide which alternative to adopt internally.

Similarly, marketing and sales would require some setting up as well. Once again this, like all the other new activities require at the practical level both an investment in the HR aspect of the activity as well as at some infrastructure. In marketing and sales, the infrastructure may also include some of the products to be used in demonstrations, for publication or other purposes, and therefore the interlinks among the different activities are more complex than just the financial links.

Among the activities that need to be considered are the identification and contracting of the different marketing channels, the logistics chain, and the materials to assist. The HR requires training, at least regarding the new products, but in some cases also in the field of marketing and sales itself. This aspect can be even more complicated in the marketing and sales are in foreign countries and especially if in different foreign countries. One may argue that for a firm to sell in both Germany and Austria, may not be more complex than in a single country, but even here, regulations regarding working hours, workers' unions and others may have some influence on the setting up of the sales force. Of course, there may be different marketing channels in the different countries, depending on the sector.

In the range of marketing and sales one should not forget or neglect the need for post sales service. This may be done remotely using a client service by phone (e.g., software) or may requires actual units that reach the customer and provide services at the clients' premises (e.g., washing machines). In each case, and there are of course middle cases and additional options, this requires setting up the right infrastructure and the training of the HR involved. Like in manufacturing, most of this, at least in some cases may be outsourced. When outsourcing there are still demands on the quality of the service offered and general monitoring. This is not the place to go over each and every potential activity, but just to demonstrate that the setting up of the activity requires investment in capital and HR including the training and the identification of the relevant infrastructure. The process of setting up all these activities requires managerial capabilities – as we discuss below in the next section.

Managerial Domain

The managerial domain regarding the required capabilities needs to be divided for the purposes of our discussion into two different stages and into two different levels.

Figure 1. Managerial Domain

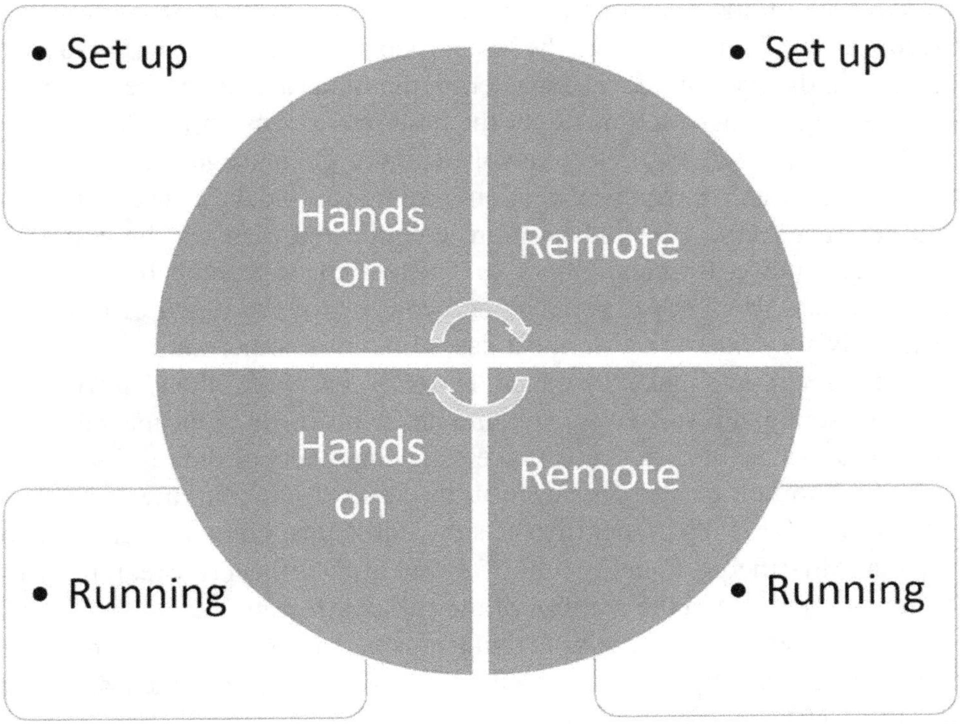

The two stages are the Set-up stage and the Running stage. In the Set-up stage the required management capabilities is the ability to identify the required infrastructure and HR resources and acquire them, including the construction of the processes in the future to set up the new activity.

While even that, can be at least partly, outsourced to sub-contractors, the setting up would still require the involvement of the management in the process to identify and acquire the advisers, and then to solve any problems related to the products, discovered when setting up the production line. The requirement for the involvement of the management in the process is fundamental as these activities will determine the future of the company and are therefore very important. The responsibility of the management is to make sure that the best available options (considering the financial implications of course) are considered, and that the company will have the ability to grow.

While this stage is very important, it is also often limited in time. The company in most cases would be obliged, either by the market (e.g., products to be marketed for the holiday season), or by obligations to investors, regulators etc.

The time limitation makes the stage a bit hectic as there are conflicting driving considerations, getting the best possible solution to allow the firm to perform at the optimum level considering the funds available; the second is completing the stage on time as per the time restrictions imposed.

The Set-up stage requires the involvement of the management and often the experience and the right connections required to find the right consultants and contract them. This can be more complicated than can be expected and is certainly a formidable task for inexperienced technically educated entrepreneurs. Here the experience and contacts of the board of directors or the advisory board, and in most cases both, are important assets to the company seeking to make the transit. However, except for the identification and contracting, this stage would require the involvement of the management in the process itself, as decision makers and approvers of the results.

It is important to note that even in the case of out-sourcing, this stage requires similar skills to identify the sub-contractors, contract them, and to make sure that there is a quality control system in place fitting the requirements of the production and the abilities of the sub-contractors.

The out-sourcing has several advantages, smaller capital investment, more flexibility in changing production technologies, smaller management demands regarding HR etc., however, the demands on the management in general are not proportionally diminished, not at the Set-up stage.

The second stage, the Running stage is not officially limited by time. It is long term activity, and therefore requires more monitoring and understanding of trends and changes. The management of on-going activities requires patience, and data management, in order to both improve or prevent problems. This is not to be confused with steady state. There may be times when there is no change in the operation and it would seem to be a steady state, but there can also be changes.

In many cases the main difference is that there is no setting up of new operations, or processes, but rather a continuation of the activities at varying levels. There can be more intensive activity, for example more production, but without the setting up of new processes or activities. That is more related to the type of activity and type of process than to the quantity or size. The production may double without any additional setting up, even if more machines are purchased. However, a move from outsourcing to internal production can be considered as a set up activity.

Of course, all of the above regarding growth would apply also to a downsizing of activities.

The two levels mentioned above in Figure 1 - Managerial Domain, refer to the type of management and involvement of the management in the operation. It can be either a hands-on involvement, in which the managers are involved in every aspect of the activity, monitoring it, and are ready to help and solve problems themselves; while the other is a more remote type of managers, that is done through delegation of responsibility and resources, and that means that there are well developed reporting and monitoring procedures and infra-structure.

The hands-on type of management is more prevalent in the TSU and is also closely linked to the setting up phase. In both cases the rapid need to for decision making, in most cases under uncertainty and with critical potential results fits the needs of a TSU.

In a complete firm, especially one that has gone beyond a certain size (e.g., over 20 people), the ability to manage and keep track of decisions and operations is diminished. That is also true for the running phase in which rapid decision making is less required, but the ability to track trends and to identify them on time, and deal with them is more important. Therefore, the remote and running are more prevalent in the complete firm environment.

However, one should remember that the move from one phase to the other, and from one level to the other may be performed independently, and in some cases, is required to do so.

Regarding the human aspect, most managers can adapt better to one of the two levels, either to the hands-on or to the remote type, and find working in another level frustrating, in the long run.

In general, the move from the TSU to the complete firm can be said to be a move from a hands-on to remote modes of management, while for the first part, there is a setting of new units and operations while later there is a move towards long term running mode of management.

Entrepreneurs, normally more at home with the hands-on management style and the setting up type of decisions required at the TSU life cycle, are therefore less at home in the new management environment of remote and running. Some adapt and some as mentioned above, leave in order to revert to the type of organization and management they feel comfortable at.

Requirements: Financial Aid; Non-Financial Aid

Accepting the benefit that the economy would derive from the TSU, and especially from the progress from a research organization into a full company employing the full range of employees, the need for aiding the Tau should be clear. As mentioned above the TSU has developed mainly the R&D capabilities internally and while outsourcing is possible, it becomes not economically sustainable after a while. The move to a full company would require not only finance for the infrastructural and growth investment required but also would require developing new capabilities inside the organization – including managerial capabilities.

The aid a TSU requires, is not always limited to the financial aspects. It is important to recall that there are normally several scarce resources that the TSU may require. The challenges the TSU faces are several. It is important to keep in mind, that the point of view of the TSU is to reach the next fund-raising stage, as well prepared as possible. The TSU will therefore need to show good performance on several plains.

The TSU needs funding, and most support programs or tools, contain a financial support element. However, since some aspects are non-reversible, like set up costs, or lead time for infrastructure ordering, the TSU in many cases faces also a time shortage. And money cannot always buy you more time.

Time scarcity is an important aspect for a TSU. This critical resource is often underestimated. The TSU needs to either reach the market before the competition, reach investors before it runs out of funds etc.

Since some activities like those mentioned above, cannot be speeded with more funds. Consider a pregnancy as an example – it takes a woman 9 months to reach a birth, but it cannot be done by 9 women in one month, and therefore not all lost time can be recuperated by funds. Therefore, one of the aspects of saving time is the knowledge how to do things, and more important, how to do them right the first time. This time spent on learning how to do things, can be saved by mentoring from experienced managers, entrepreneurs. Support programs that supply this demand are very crucial.

Example can be seen in the mentoring services the EU provides to SMEs that have been funded. They are eligible to get consulting services from the EEN network, that will allow the managers of the small firm access to knowledge they would have difficulty in accessing otherwise.

In the TSU environment such services can be critical not in the technical domain, as there the founders are normally the experts and thus do not require help, but rather in the more business-oriented activities. Marketing, finance, IP issues and similar activities are often a more pressing need for the TSU managers, more so, as they normally lack the private connection to supply them with the needed advice in these fields.

It is important to remember that outsourcing solves part of the problem but not the entire issue. As an illustration of this point consider IP protection. The technical founders of the TSU would not submit their patent, as they know better and would approach a patent attorney. However, they still need to be able to understand the different terms, the process of application and examination, the contribution to the timeline of PCT and other international instruments, and finally the cost of the activity. Even if an activity is outsourced, in most cases the TSU would still need to be able to manage it and the sub-contractors taking care of the task.

Government aid in Market Entry

In this section we shall discuss the specific requirements for market entry, that the TSU maturing to a complete company has to face. The limitations on the government aid that is available for such markets, and the aid that is more readily available through conventional means.

Market entry in a complex issue. In order to make it successful, a good understanding of the specific market segment is required, as well as a very good knowledge of the market players and the rules. This is a significant challenge for any external player, and one that is not trivial even for players in the market.

Markets are changing, players enter and leave the market, external influences from potential players, as well as regulators, impact the current and potential activities making the understanding of the market an on-going effort. For entities external to the market, the effort required to really understand the market, is even bigger. It is not enough to buy a general market analysis, or even to try and order a specially tailored one. The demands are expansive, and for a new entry, the biggest question, is often which market?

There are international entities as well as governmental entities that can help. The chambers of commerce associations in many countries can help, especially with introductions to lead consultants, or the contacts that can assist in market entry in the target countries.

Diplomatic missions often include a commercial attaché, with a specific unit designed to assist the country-of-origin entities in entering the target country. This is one of the ways in which a country can assist in a market entry. Most countries would actively encourage the ability of a firm to export as this brings foreign funds to the country and helps the local economy.

Regarding market entry in the local, national market, here things are more complex. Entering the local market could normally come at the expense of one of the current players. True, in some cases the market entry may change the rules of the game and make the market bigger, but even in the cases when this happens, it entails a relatively long period of time, and at the beginning, at least one of the current players pays the price, until the market grows and it can recuperate, if at all. In most cases therefore, any assistance in local market entry will come at the expanse of another player and this creates a conflict for the government. The government cannot prefer one local player to another without being accused of corruption. There can be cases when the current players are all foreign, where such an intervention on behalf of a local player can be justified, but that is rare. One significant case for the involvement of government in new player's market entry is the increase in competition. Increased competition is considered to be a good thing for the client, often the citizens and thus a good thing for the government to encourage.

Example: *a group of entrepreneurs wishes to fund a new bank in a certain country. The current banks in the country demand high rates of service costs, and thus the government consider the entry of a new player, ready to reduce the transaction costs for the clients a good idea. They offer the new group faster regulatory approval process. They invite in fact also foreign banks to open branches and offer services in the country. The reduction in the transaction fees, not only increases the amount of transaction, but by doing so increases the financial activity in the market and creates growth nationally.*

Therefore, most governments would not have ready tools for local market entry support, unless they can show that it increases the competition in the market, and that they offer it to all local (or even international in some cases) entrants without special favors.

We have seen up until this point company specific measures that governments employ, but there are also more general measures that governments can employ to facilitate market entry or to raise barriers to it.

General Government Assistance

Government assistance is not limited to financial aspects alone. It can also be in the regulatory aspects, which are in the government control. Simplification of the regulatory demands can easily reduce the market entry barriers and is certainly within the government control.

A closely similar area of support is standardization. While not always a governmental tool, standardization is in many cases a public activity and as such is commonly a governmental issue, allowing again the place difficult or relatively small barriers to market entry, as per the government wishes. Standardization simplification can be made easy for example by having the standards published in English as well as the national language, thus allowing foreign players easier access to the standards. Adoption of international standards is another such measures. There are of course the standards themselves which can be made easier to comply, as well as the certifying bodies which can be more, or less, helpful in certifying compliance. It is important to recall that standardization is also in some cases a protective tool for the local economy against foreign players, and therefore changing that policy could entail some resentment, which having potential political implication, could influence the government.

Duties and custom on imports, are an important market entry barrier. In most countries these demands have been placed as a protective measure for the local players against foreign ones. Therefore, governments are active in bi and multinational trading agreements which aim to remove such barriers and allow for a market entry to foreign players from a certain country.

Specific Government Assistance

A specific assistance to a company would be focused on a specific company rather than a group of companies, a policy, or a sector. In this case the government is limited in its ability to support a specific company. Supporting a specific company would require the government to demonstrate a special case, otherwise, the move would be deemed to be corruption. In this section when we deal with specific assistance to a firm, we shall refer to assistance given to firms requesting it on a basis of a program, or a special assistive rout. Setting pre-arranged public and published guidelines and criteria.

A specific assistance can be done via setting a special program for lunching new products into the market, by supporting market entry via assistance in marketing (funding for marketing activity) or similar ways, which are granted to specific applicant companies, according to pre-defined criteria and awarded based on that criterion alone. Examples of such routs or programs are presented below.

A generally known example of such a support program can be the SME Instrument program operated under the Horizon 2020 program. In the SME Instrument program support is offered to SMEs in different stages of product development in order to promote them towards the market. The criteria for awarding the funds at each stage, there are three stages in the program, but only two entail financial support, are clear cut and consistent over time. It is therefore clear, that while the support is given to individual firms, the decision is based on clear criteria, announced at the start of Horizon 2020, and thus that there is no special treatment for specific firms. It is clear that the program was designed to assist SMEs with the best potential assessed for the future, and not tailored made for a specific SME.

The two main elements that eliminate the claim of subsidizing specific firms is that the criteria for awarding was declared at the start of the seven years' program and thus there could not be any suspicion of a specific aid; that the criteria were scrutinized by several entities (EU parliament, the commission, and others) and that the criteria were approved by all.

Therefore, either criteria that are criticized since publication, but are still enforced, or criteria that are announced after the award, or submission, or that not published at all, give in all these cases, a cause for suspicion of underhand behavior.

Another possibility to at least reduce the potential criticism, is to demand a return of the grant (making it a loan) after a certain project period. This reduces the suspicion as there is no granting of funds, but rather a reduction of the funding cost, and that would in many cases not warrant underhand behavior, in the public eye.

Ecosystem: Supporting New Entrants; Opposing Them; Access to Market

The ecosystem is comprised of many types of entities and is characterized by a unique character that allows to identify it specifically. The SUs or TSUs ecosystem will include the relevant (probably geographically) TSUs, their real-estate suppliers of different characters, the relevant financial institutions, as well as governmental and municipal entities that are involved in the activity.

The ecosystem will also include different consulting bodies, from marketing, legal and accounting firms, to design, production, strategy etc., type of consultants and free lancers. The ecosystem may include among other, veteran entrepreneurs that serve as mentors and board members to the TSUs helping them by advice and connections.

The ecosystem will create the environment of mutual support, the encounters between the TSU, the government agencies, the consulting entities, and the financing agencies, that are needed to help promote the TSUs along the way.

There are therefore support tools that can assist from networking and networking tools to assist in networking between the different entities in the ecosystem, to mentoring platforms, and direct learning and education platforms that will help the entrepreneurs in their learning challenges.

This could start as early in the TSU life as encouraging people to lectures given by experienced entrepreneurs, in order them to establish their own startups. Combining this with meetings with angel investors and assisting in the formation of the TSUs and all along the TSU life, leading to teaching the basics of M&A, encouraging meetings to allow such events, workshops fairs etc.

In fact, the more diverse, complex, and active the ecosystem, the more attractive it will be for the TSUs and thus it will attract investors, and the other members, in a positive feed-back mechanism.

Examples of Tools: Beta Site Support; National R&D Institutes; Production Labs

In many R&D products there is a need to present a prototype and have it examined by the potential clients and the market. In fact, this is one of the TRLs.

However, while the stage may be well known and acknowledged, it is difficult to find funding for this stage, and many times this is a crucial stage. There are not many examples for such a support, but the SME Instrument in phase two actually assists in overcoming the stage of presenting a prototype in the final working environment. The program allows the grant receiver to prepare and present such units and to test their operation.

In other cases, governments have set up special units that can help the industry solve production problems, or development issues, and these, national R&D institutes, which are partly supported by the government, are in fact, subsidized help. The setting up and a significant portion of the operational costs of these institutes, are covered by public funding, and they allow the companies to get the help they need, without sharing the setting up costs. It is a much subtler way of supporting the firms, and it gives the same support to a small as well as a large firm.

Production labs, or access to existing research infrastructure, as in Horizon 2020, is another subtle example of such help. The setting up and some of the operation is nationally supported and thus presents another possibility for assisting specific firms by using a general tool. These infrastructure installations are even paid for the operation, partly, in a participatory way by partially funding the access itself. That support is given in addition to the original setting cost of the infrastructure.

FUTURE RESEARCH DIRECTIONS

The topic of government support and government support tools and programs is widely researched (Porath 2017). The questions of how, when and how to control government support would continue to interest academic and practitioners in the future.

CONCLUSION

In this chapter we have discussed the assistance that can be given the TSU in the early stages.

We started the discussion with a general overview of the aspects of government support. We discussed them briefly before moving on to a discussion on the nature of market failure and the reasons for need for help. We mentioned briefly the nature of government support schemes.

We moved to discuss the benefits, economic and sociology aspects of the TSUs and the importance of assisting the TSUs in order to secure these benefits.

We moved on to discuss the capabilities the TSUs need to develop and went into a detailed discussion regarding the managerial capabilities the TSUs need to develop and compared them to those of the complete company.

We differentiated next regarding the financial and the non-financial assistance, and the government role in both. We discussed the example of market entry, local and international to further illustrate the point.

Using more examples, we discussed general government support as opposed to specific governmental support.

We ended with a discussion on the ecosystem, its characteristics and the tools and assistance it can offer, comparing it to other tools and programs.

Even with all the examples and illustration, it is time to discuss in depth a real case study that will illustrate the move that we have analyzed so far, mainly theoretically. This will happen in the next chapter.

REFERENCES

EU. (2000). *Lisbon European Council 23 and 24 March 2000.* https://www.europarl.europa.eu/summits/lis1_en.htm#1

EU. (2003). *Communication from the Commission; Investing in research: An action plan for Europe.* COM (2003) 226 final. https://ec.europa.eu/invest-in-research/pdf/226/en.pdf

Luukkonen, T. (1988). The difficulties in assessing the impact of EU Framework Programs. *Research Policy*, *27*(6), 599–610. doi:10.1016/S0048-7333(98)00058-4

Luukkonen, T. (2000). Additionality of EU Framework Programs. *Research Policy*, *29*(6), 711–724. doi:10.1016/S0048-7333(99)00041-4

Porath, A. (2010). *Collaborative Research in Management: Inside out*. SAGE Publications.

Samila, S., & Sorenson, O. (2011). Venture capital, entrepreneurship, and economic growth. *The Review of Economics and Statistics*, *93*(1), 338–349. doi:10.1162/REST_a_00066

Storey, D. J. (1991). The Birth of New Firms – Does Unemployment Matter? A Review of the Evidence. *Small Business Economics*, *3*(3), 167–178. doi:10.1007/BF00400022

Taleb, N. N. (2012). *Antifragile. Things that gain from disorder.* Random House Trade Paperbacks.

KEY TERMS AND DEFINITIONS

Beta-Site Testing: This is an advanced prototype testing that is carried out at a potential client site, or otherwise mimicking real market conditions for testing the product and its value to the clients.

Financial Aid: Support either by grant, loan or loan guarantee of funds aimed at assisting (in our case) the TSU in overcoming barriers that are not well supported by private market support.

Market Entry: The introduction into the market of a new product or technology. This is the first time the specific product or technology (of the specific enterprise) is presented to the market (market segment) and this would determine in most cases how the product will be incorporated by the market, and the market share it will acquire finally. This is a strategic action and requires both planning and market understanding as well as risk management.

Market Failure: A lack of positive action from the private market, or private entities to support a commercial activity. This is often due to perceived high risk, or lack of clear benefits resulting to the private entities involved. It means that certain activities are not supported by private entities even if these activities would benefit the economy in general. In that case it is often up-to public (government) entities to get involved and make supporting the activities more attractive to the private sector or support the activities directly. This is done in order to benefit the economy in general.

Non-Financial Aid: A similarly aimed support which is not financial but can take other forms (e.g., mentoring support, general lenient for TSU regulation).

ENDNOTES

1. There are several gaps noted for the R&D process, impacting the performance of a TSU.
2. For a definition of market failure please see: https://www.econlib.org/library/Topics/College/marketfailures.html

Chapter 8
SeeTRee Ltd.
A Case Study

ABSTRACT

In this chapter, the authors present a special case study of a company that has managed to evolve from R&D activity to a complete company. The case will describe the special story of the company and the specific characteristics and changes the company has undergone. The case study deals with the move from an R&D-oriented company, developing its basic technology, service, and business model, towards an income generating firm. As much as possible, the technology and related technical terms are omitted from the discussion both to keep it clear and focus on the narrative, as well as to avoid undue disclosure of IP.

DOI: 10.4018/978-1-7998-5685-6.ch008

GENERAL INFORMATION

See Tree Ltd. Is an Israel start-up company founded in August 2017 by three partners Israel Talpaz, Guy Morgenstern, and Barak Hachamov. They represent a combination of talent and experience of military intelligence, engineering, and high-tech entrepreneurship. They raised 3.2 million US$ with pre-investment company value of 5 million US$ in August 2017 and launched the company. Since than they raised a round A of investment of 11.5 million US$ with post investment value of 35 million US$ in 2019 and are now planning round B.

The company is located in the Tel-Aviv area and currently employees 60 workers.

The company goal as defined to its potential investors[1] is: To optimize fruit production, tree health, and transform permanent-crop farming. We do this by providing intelligence- per every single tree, for many millions of trees, at a cost of about 10 cents per tree per year.

All three co-founders are involved with the daily operations of the company and its development.

The Business

The SeeTree Ltd. ("**ST**") (2020) identified the following need[2]: "Farmers today, simply put; do not have the quality of data they need to enable the optimization of their fruit production. They still rely on their legs to collect the data, on their eyes to assess the data, and on their intuition and past experiences to make decisions based on the data. Their decision-making process relies on data that lacks accuracy, consistency, comprehensiveness and scalability."

It takes technologies that are used for other purposes in other sectors and translate them to the agricultural sector, with specific focus on trees and plantations, offering a service under Precise Agriculture.

The company identified that while in other crops, field crops, cereals, and live-stock management there was some advancement in precise agriculture, and the services offered to farmers were improving, the same could not be said for trees and plantations.

They offer a service in which they run a data collection and analysis campaign that is multi layered and is based on multiple sources. They collect data and create a data set in which each tree has a dossier including different data about it, allowing them to offer analysis per tree. They use their collection and analysis capabilities to offer not just data about specific trees, but their ever-growing data set allows them to make horizontal analysis and present insights, that could not be presented before. We shall present such examples in the sections below.

The ST Short History

When Israel retired from the intelligence business, he was looking to utilize his expertise and the technologies in new sectors. Basing his interests on the agriculture family tendencies he turned this way and identified the need for smart tools for precision agriculture in permanent crops. Thus, following six months of preparations and seed funds raising, with his two co-founders, Guy Morgenstern, and Barak Hachamov, founded SeeTree Ltd., in August 2017.

The company started working like any other technology-based start-up, by doing R&D, but following three months of in-door development, in November 2017 reached out to its first in the field client, a farmer in California, to start doing some field tests.

The company since then, has been developing its products and services, based on the constant interaction with the potential clients. The systems developed, and the services offered work in the following way: they start with a field test, for free, to show the potential client what they can do for them on the clients' own turf and own crops.

This free trial allows ST to develop their data set, specifically designed for the specific conditions of the client, and to train their system, on that specific dataset. ST claims that this allowed them to better train their AI and thus to achieve exemplary results, allowing them to conclude deals. The free trial allows them to develop the data set without pressure from the client / farmer, who does not press them for results, as they cost him nothing.

This did not go well at the beginning with the investors and especially with the more conservative VC type investors. They wanted ST to either start charging payment for the trials or encouraged them to sit in the offices and work on the further development of the products and services, and not to "lose focus" by getting involved in projects in the market.

ST maintains that the free trials, without pressure and the constant contact with the clients and their needs, allowed them to develop the set of tools and services that attract the clients the most. It actually made their product more relevant, than if they had sat in the office away from the market and continued to develop the set of tools, based on the imagined interest of the clients. ST recruited as CSO Ori Schachar a 13 years' veteran of Mobileye, with a unique understanding of creating value in AI systems[3].

The company at this stage had an organizational structure typical of most start-ups, everybody is involved with every aspect of the work. Little specialization in the activities and even less importance of the official titles. This changes about a year and a half later, when coinciding with round A of fund raising, a different structure was established.

The new structure is that of regional desks, California, Brazil etc., led by a customer satisfaction manager, dealing with the local projects and clients, and being supported by the R&D unit, and the product unit. The desks are all under the COO, and operate directly with the clients, doing the free trials and moving to the services, where they are being supported first by the R&D and later by the product units. This structure allows the firm to expand into new markets, by duplicating the desks for new regions and getting the same support from the product and from the R&D units.

Round A was intended to raise 6 million US$ with a pre-investment value of 18 million, but demand was so big that an additional 5.5 million US$ were raised, and pre-valued at 30 million US$. Closing the entire round with 11.5 million US$ at post investment value of 35 million US$.

The boost in the investment interest lies among other reasons in the fact that the new start-up was already generating more and more income with a list of ever-growing satisfied customers. This was the best proof of the business potential for the investors that any start-up could offer.

The company is now focusing on expanding the business into new territories, to be able take hold of as much as possible of the potential market and increase the number of trees under ST management.

Each tree is allocated a dossier, which contains not only current data about its growth and yield but also a historical data about its performance in different times along the years.

The data base of trees which today counts about 30 million trees, allows the company to identify trends that could not be identified before, and to give a better service to the clients. Israel remembers that when they first offered their services to orange growers in California and asked the farmers if the height of the canopy was interesting, the farmers said "NO". Thus, they removed that parameter from the services in California. When they entered the Brazilian market and presented the services to the clients there, height and size of canopy were interesting as the shadow of one canopy could influence the fruits in a neighboring tree. ST could prove that using its tree dossiers and with that proof Israel went back to California which now was far more interested in the canopy size and height.

ST combines a multi layered data collection and analysis campaign that is unique to them in comparison to other companies operating drones and other aerial platforms. The combined proprietary HI and AI capabilities of the firm allows it to create a health map of the entire plantation, detailing every tree with a dossier of itself. It allows the farmers to identify the value of each tree specifically.

The company thus is also able to detect HLB from the air in citrus trees, a disease that has no known cure, and which caused Florida to lose 80% of its citrus production. The company created new indices defining the tree health level and can use this measure to evaluate the need of the trees and the value of each tree, regarding growth and yield.

The company can also assess the volume of the fruits and their size, before the harvest, allowing for better planning and reducing operation costs.

This combination of a large data set allowing to migrate lessons from one place to another, is important, just as the specific dossier of each tree is important for companies targeting watering, fertilizing and counter disease spraying of individual trees, in precision agriculture. This allows the interested companies to improve their services and product.

The academy is also interested in cooperating with the company, as its unique data set offers special possibilities for research.

The company understands that in order to remain attractive and to become even more so, it should expand fast. The B round now being planned is designed to allow to invest in rapid expansion, and in automation of the products and services even further. It is planned to raise 20 million US$ with pre-investment value of 50 million US$. Following the B round the company will raise the C round after two more years and will reach the breakeven point in 2023.

The next organizational changes will take place within the R&D, where further specialization is required to allow for improved R&D. The regional desks will stay as they are, at least up to the break-even point.

CONCLUSION

There are different lessons to be learnt from this case study in market penetration, lessons related to investor relations and strategy, product development etc., however, in this chapter we shall discuss the lessons relevant to our topic, the move from R&D to the market.

The structure of the company at the beginning was typical to that of the TSU at its R&D phase. Everyone, certainly the founders are involved directly in all the activities, leading them, and personally overseeing the progress.

However, once the company has grown in activities, number and complexity and started collecting income, it has changed its structure to be more market oriented, giving more attention to the market and its demand, while the R&D was relegated to a support role. The market was identified as the leading engine of the company growth, and while it is still not financially independent and requires further investments, it has certainly acquired a more professional and hierarchal structure. The new structure is designed to allow the firm to expand its activities in the market, not in the R&D. The desks, already comprise other disciplines than R&D and in the future will include a dedicated to the region of the desk – sales force that will be tasked with increasing the volume of the sales.

This new move will complete the transfer of the company from a TSU to a complete firm.

ST is an unusual case, one that demonstrates a smooth move from the R&D phase to the complete firm phase, mainly due to the stage at which the move was made.

Since the start, three months after seed investment and firm establishment, the company moved to an R&D mixed with market involvement model. It had to develop the "complete firm" capabilities in order to do that. So, when following A round they changed the organizational structure, the activities remained the same, just the structure and job definition changed. The "shock" was reduced, as there were no new activities to digest and new management capabilities to develop. There was the re-organization and that was a shock by itself, but it was minor compared to companies who have to deal with new organization combined with new capabilities, new activities, and new types of people. One of the results of the special strategy was that when the new organizational structure was formed, the desk organization, no workers left the company. As Israel described it, there was some friction due the change in the balance within the firm, but in the end, all the workers stayed and helped the move. This is an important factor to recall when considering the strategy of similar firms.

The new organization is once again a combination of small groups who with the support of the R&D and the product teams are in fact small TSU's limited to a single region or project and thus maintaining the TSU spirit. The next expected change in the R&D structure may be more problematic.

However, as there could be bigger changes in the future, it is possible that this is not the end of the maturing process for the firm. For this point, the changes have proved both useful and less dramatic than could be expected due to the two points: maintaining a complete or almost complete set of activities and the relevant personnel almost from the start; and keep the spirit and organizational culture of the TSU in the regional teams.

It is important to remember that just as Israel mentioned, most TSUs do not follow that example, but rather develop away from the market, in order to create value as fast as possible; to maintain focus, and reach the market asap, before the competition. These are also the advice ST got from their investors. The investors wanted to avoid the difficult lessons from the past where firms getting close to the market, are confused by the myriad of opportunities, start to chase side projects, and lose both the focus and their leading role in the market.

It is also important to recall that the main interest of VC and of many entrepreneurs today, is to reach the next investment round at the best position, to raise funds with the best pre-investment valuation, the market in many cases is considered far away, as it is for most TSUs beyond the next round of investment.

These interests drive most TSUs, and here ST is an exception, to focus entirely on the R&D and not to dedicate precious resources, human and financial as well as managerial share of mind, to develop active marketing, customer satisfaction in the ST case, or customer service in other etc., thus making the move from TSU to complete firm more complicated.

In getting early to a semblance of a complete firm, while the TSU culture is still prevalent, before round A, ST managed to allow the management to build the country desks, the R&D and the Product units from the original HR they had in the firm, and from there to have mostly an organic growth. It also prevented the cultural shock resulting from the introduction of new disciplines into the firm and into the managerial capabilities roaster, as these were already in the company, just organized differently. The fact that not all the HR was R&D oriented, or even technologically unified, allowed a smother transit.

REFERENCE

SeeTree Ltd. (2020). *One Pager for Customers.* Author.

ENDNOTES

1 Taken from the company letter to the investors.
2 There.
3 Personnel mentioned held the ascribed positions in 2019.

Chapter 9
The Learning Process as a Maturing Process for TSUs

ABSTRACT

In this chapter, the topic of the on-going learning, especially in SUs and specifically in TSUs, is discussed. The move from a TSU to the next stage discussed in length earlier from the practical aspects is discussed here as part of the learning process of the TSUs. The learning process in TSUs is presented. The early stages of learning to manage projects and the project results are presented as part of the TSU core activity in its role as TSU. The next logical step is to discuss the move to a complete mature firm as another stage in the learning process of the TSU, as a step characterized by learning and enhancing the managerial capabilities of the firm. Some of the research relevant for learning in TSUs is presented as well as the specific case of TSUs inside a technology incubator.

DOI: 10.4018/978-1-7998-5685-6.ch009

INTRODUCTION

The discussion for far focused on the practical aspects, the barriers and assistive tools, the required changes and mainly the difference between the firm as focused on R&D and innovation versus the developed entity dealing with all aspects of the active firm.

However, the move from the one to the latter is not necessarily a novel action. It can be viewed as part of the learning process the TSU undergoes from the ideation stage to the readiness for the move into a complete firm. The learning process described below is a well-researched topic in management as the realization that entrepreneurs and the entity they are forming has to learn to cope with new challenges all the time (Calantone, Cavusgil, and Zhao, 2002; Delmar, and Shane, 2006; Hughes, and Morgan, 2007; Tsegay, Shi, and Wu, 2016; Entrialgo, and Iglesias, 2017; Yin, and Lou, 2018; Gabai, Ehrlich, Abdel Razek, and Krieger, 2020). There are differences between the stages where the learning has to be done by the individual and when the learning is done by the "firm[1]". There is of course an important aspect of learning not only to perform the tasks but also in learning how to manage them, creating managerial capabilities in the firm. In fact, when dealing with out-sourcing, the aspect of managerial capabilities is paramount.

It is therefore important, once the move discussed in earlier chapter and its challenges are understood, to see it also from a different perspective, that of a learning move. In this chapter this will be the main discussion point.

ENTREPRENEURIAL LEARNING

Facing the Challenges

Recent research into the most effective learning methodologies, indicates that the best alternative is to develop a culture of learning in STUs by acquiring as soon as possible Organizational learning practices (Baltrunaite and Sekliuckiene, 2020). The importance of having learning processes should be self-evident, however, just to make sure it is clearly understood let us review it.

The early stages of a new enterprise are shadowed by performing activities for the first time. Even when the teams are well experience, unless the entire team has already performed these activities in another venture, there is some newness in the activity. There is a need not only to perform new tasks that have never been done before by the new entity being formed, but also to learn how to liaise with others in the organization, how to manage the processes of activity and reporting, and how to improve these towards the future.

Simple tasks, such as recruiting HR (not simple at all but in fast growing organizations, a daily event), ordering equipment and materials and others, require the formation of approval process, as well as monitoring and reporting, and that on top of the professional aspects of the activity – how to recruit, how to identify suppliers, engage them, make the order, check the supply and pay.

The enterprise learns how to perform the new activities, and does so in an ever-changing environment, both on the inside as well as on the outside. It is important to remember that tasks, such as project management, controlling the products of the projects and incorporating them into the firm are not as trivial as it would seem from the outside.

Due both to their relative rareness (less than HR recruiting and purchase ordering) as well as their importance to the value creating in the TSU, these are crucial actions to be learnt as well as possible.

One of the roles of different support entities such as technology incubators or accelerators, where they house the new enterprises at their earliest stages allowing them to focus on R&D while giving the administrative umbrella, may assist the TSU by reducing the learning load – no need to learn the full purchase activity and some of the other administrative actions, but does not relieve the need to learn to manage the R&D project and the adoption of its results.

We have discussed before that some of the actions can be outsourced (the incubator is a way of outsourcing them, but not the only one). However, in our discussion before we mentioned that even the out-sourced actions require some resources and especially managerial capabilities and share of mind as well as time. In this chapter we need to understand, that in addition to that, the outsourced activities require first – learning. Learning how to contract them (searching, identifying, contracting the out-sourcing contractors, and managing and monitoring them and the quality of their work), and later on, developing the internal processes of monitoring and reporting on that activity.

The learning effort of the out-sourced actions is perhaps reduced compared to in-house actions, but it is there and should not be forgotten.

Learning and Firm Development

The growth of the firm from a TSU to a complete firm can be viewed as an increase in the technical and managerial capabilities of the firm. With a special emphasis on the managerial capabilities embedded in the management.

The accumulation of capabilities evident not only in the specific experience of the personnel involved in the company (transient capability), but also in the procedures and customs of the firm. A firm that has learnt at first to manage R&D projects, will have process and customs regarding deciding on projects, financing them and the required reporting and approvals along the way. Of course, the personal experience of the personnel involved is important, but that can be viewed as a temporary or Transient Capability. With the high turn-over of HR in high-tech companies and in TSUs, the experience of the people involved is not enough to create a real capability.

The inherent process and customs, state clearly that there are certain capabilities, divorced from the specific experience of certain personnel in the firm.

For example, if a TSU recruits its personnel based on the personnel networks of the current employees, this ability and the type of people recruited would change if people leave the company and others replace them. If at the beginning a group of people were recruited from a certain technological college, they would naturally tend to bring people from their college and around their age. If a group of them decided to move to another firm and other people came in from a different college, they would tend to bring their friends from this new college. If, however, the company decided to source out the recruitment to an HR company it would introduce employees of greater variety of colleges and ages.

Following the above-mentioned example, if the company at first outsourced the recruitment, it would develop process to decide upon recruiting and the type required, perhaps also regarding the selection of the HR firm. If at a later stage the company as part of its growth, decided to create an HR department that would also deal with recruiting, it would develop processes for searching for employees, processes for evaluating candidates etc., that did not exist in the company before. These new processes and the approvals accompanying them would create long term capabilities in the company.

TSU's as Noy and Ellis (2003a; 2003b) noted, are risk averse in areas not relating to their technology. Therefore, they TSUs tend to be very conservative in areas of management not related to the technology. This leads to two characteristics of the learning. At first the TSU would be conservative and either follow the guidelines of the guiding organization (technology incubator, technology park etc.), if it has one, or the example of other TSUs it has in its network. As the TSU growths and experts of new activities are recruited into the TSU, one would see a break from the guidelines, as these experts are ready to take risks and not follow the guidance of the guiding organizations or seek better value for money solutions than those employed by other TSUs in the network. Of course, a TSU that is merged into another company does not face these challenges.

TSUs in a Technology Incubator

The realization that TSUs need more than just funds, e.g., mentoring was recognized in some places earlier on. One such example is the Israeli technological incubator program, operated since 1990 by the Israeli Innovation Authority (IIA)[2]. The technology incubator program funds projects housed inside the incubator, which is a company giving the supported firms it hosts, administrative services as well as others allowing the firms to focus on value creation via R&D. The IIA supports the firms with a conditional loan they have to repay if successful, and the complementary funding is either raised from private sources or from the incubator itself. While the IIA does not take any share of the supported firms, the incubator does take equity in the firms in return for its services as well as complementary funding. The program has been revised several times over the years, to adapt to the changing ecosystem.

When discussing the SU's above the focus was on TSUs, and especially those that are managed as independent firms focusing on R&D. SUs, business or technological that are spending time in an incubator are a special case (Israeli Innovation Authority, 2022). The nature of the incubator is that it allows the firm to focus on maturing a specific aspect, in order to allow it to raise funds in the "real world". There are also private incubators without government support, where the support of the incubator is dependent upon the model of the specific incubator, and there are also business incubators. The business incubators are set to allow the firms to validate their business model (market testing, preliminary sales), and thus to validate their viability, increase in value and prepare for the next investment – at a higher value.

In a Technological incubator the TSUs are expected to reach a PoC that shows the viability of their technology and product and that allows them to decrease the level of uncertainty. That uncertainty is technological mostly but can also be business and regulatory. That is done for a similar purpose as the business incubators – better position for future investment.

Just the description above, would generate the understanding that in order to allow the firms the special focus required in the incubator, would mean that the incubator will take on some of the aspects that may diverge the focus – financial management, HR and others. This would mean, that in the case of an incubator the firms housed there, would probably be even less "complete" than TSUs operating independently, helped by sub-contractors. That would create a "crisis" for the firms at the end of the incubation period when they would need to leave the support services behind and manage on their own. This specific case would require further investigation into the difficulties these firms face.

FUTURE RESEARCH DIRECTIONS

The learning of managers and especially of entrepreneur will certainly attract further research, both regarding models for such learning and the way it is done, as well as regarding case studies showing how successful managers / entrepreneurs do it.

From the other topics, a case of understanding the challenges firms leaving, after graduation the support system, accelerator, incubator and similar would require further research. The challenge as I see it is in understanding if the focus the firms are encouraged to embrace while neglecting the development of peripheral capabilities (the ones supplied by the support), is beneficial to the firms, or hampers their development.

While the focus allows them to diminish the uncertainty, and thus perhaps raise the next round of funds, being a beneficial outcome, it may also predetermine their preference towards an M&A, as they lack the capabilities they did not develop. If that is proven to be the case, it is a question for the decision makers, if this is the rout, they want the TSUs to take, or perhaps they would rather develop new tools of support.

CONCLUSION

The basis of entrepreneurship is constant learning, as the need to deal with new challenges is very urgent is TSUs. The learning process in TSUs is therefore crucial for the understanding also of the changing phases in the TSU life once it is ready to move from a mainly R&D activity into a complete firm environment. It is important therefore, in order to deal with the change, to understand the learning, and to understand the different support activities available for TSUs and the impact each such decision may have of the future track the TSU will take in the future.

This chapter appears at first look to be directed at policy makers and implementors. However, the process, the support tools and the resulting impact both may have on the preferences of the TSU, when needing to select a path are also crucial for the specific entrepreneur, selecting a track at this moment for the new firm formed. It is in any case, part of an ongoing debate rather than a settled conclusion.

REFERENCES

Baltrunaite, V., & Sekliuckiene, J. (2020). The Use of Organizational Learning Practices in Start-Ups growth. *Entrepreneurial Business and Economics Review*, *8*(1), 71–89. doi:10.15678/EBER.2020.080104

Calantone, R. J., Cavusgil, S. T., & Zhao, Y. (2002). Learning Orientation, Firm Innovation Capability, and Firm Performance. *Industrial Marketing Management*, *31*(6), 515–524. doi:10.1016/S0019-8501(01)00203-6

Delmar, F., & Shane, S. (2006). Does Experience Matter? The Effect of Founding Team Experience on the Survival and Sales of Newly Founded Ventures. *Strategic Organization*, *4*(3), 215–247. doi:10.1177/1476127006066596

Entrialgo, M., & Iglesias, V. (2017). Are the Intentions to Entrepreneurship of Men and Women Shaped Differently? The Impact of Entrepreneurial Role-Model Exposure and Entrepreneurship Education. *Entrepreneurship Research Journal*, *20170013*(1), 1–14. doi:10.1515/erj-2017-0013

Gabai, U., Ehrlich, E., Abdel Razek, K., & Krieger, N. (2020). *High-Tech Human Capital Report 2019*. Israel Innovation Authority and Start-Up Nation Central.

Hughes, M., & Morgan, R. E. (2007). Deconstructing the Relationship between Entrepreneurial Orientation and Business Performance at the Embryonic Stage of Firm Growth. *Industrial Marketing Management*, *36*(5), 651–661. doi:10.1016/j.indmarman.2006.04.003

Israeli Innovation Authority. (2022). *Technological Innovation incubators program*. Downloaded on 19 December 2022 from https://innovationisrael.org.il/en/program/incubators-incentive-program

Noy, E., & Ellis, S. (2003a). Risk: A neglected component of strategy formulation. *Journal of Managerial Psychology*, *18*(7), 691–707. doi:10.1108/02683940310502395

Noy, E., & Ellis, S. (2003b). Corporate Risk Strategy: Does it Vary Across Business Activities? *European Management Journal*, *21*(1), 119–128. doi:10.1016/S0263-2373(02)00159-7

Tsegay, B. K., Shi, C. S., & Wu, W. W. (2016). The Impact of Innovation Strategy on Organizational Learning and Innovation Performance: Do Firm Size and Ownership Type Make a Difference? *South African Journal of Industrial Engineering*, *27*(1), 125–136.

Yin, B., & Lou, J. (2018). How Do Accelerators Select Startups? Shifting Decision Criteria Across Stages. *IEEE Transactions on Engineering Management*, *65*(4), 574–589. doi:10.1109/TEM.2018.2791501

KEY TERMS AND DEFINITIONS

Israeli Innovation Authority: A corporate created by law in Israel in 2016 taking over the functions of several units within the government, and without, supporting innovation actions. It replaces the Office of the Chief Scientist of the Ministry of economic development and is operating the programs formerly operated by that office.

Technology Innovation Incubators Program: A national program supporting TSUs in technology incubators. The incubators are firms selected by tender for a certain period in which they are selecting TSUs and together applying for financial support for the TSUs from the IIA. The incubators, receive for their services, equity, as well as additional equity for supplementary financial support. This program, started in 1990 has undergone several changes and is considered an engine for the creation of TSUs.

Transient Capabilities: The capabilities that are linked directly to specific personnel, based on their experience or training, and which exists in the firm as long as the specific person is employed. When the personnel change, the capabilities will change with the people leaving and the new people replacing them. These capabilities are not reflected in the processes and customs of the firm which remain in force even with personnel change.

ENDNOTES

1 The Firm is comprised of individuals who retain the non-tangible knowledge, but here the discussion refers not only to 'how to perform a task' but also how to manage and liaise with others in the firm.

2 The IIA was formed in 2016 taking over the responsibility of the former Office of the Chief Scientist of the Ministry of economic development.

Conclusion

Summing the challenges and the potential advantages of the move including the support tools available in different countries.

The move from the relatively safe and familiar world of R&D in the TSU to the complete firm and its environment is a traumatic move. We have described the nature of the TSU and its environment and compared it to that of the complete firm.

It is clear that a company focused on R&D alone, living from investment, is very different from a complete firm, competing in the real market, with fully developed department, surviving from the sales of its products and services.

The environment in which each operates is different, therefore the organizational culture, structure and management demands are different. We have analyzed these differences and discussed also the differences stemming from the different types of employees employed by each organizational species.

It is clear, that for the founders, managing the TSU the move demands some evolvement in the managerial capabilities and the demands placed on them. They also, as the persons in charge of directing the move, have the responsibility of formulating the strategy regarding the move.

We have mentioned the two main avenues for proceeding with the move, the M&A with a complete company (normally a larger one then the TSU); or the development into a complete company by growing the relevant units - including the resources, the HR etc.

These two main strategic avenues have their own advantages and disadvantages, and as we have shown there are different aspects to be considered. There is the point of view of the founders and their interests, the culture and the organization of the TSU and their appeal. There is the financial aspect of the complete firm and other interested parties involved.

One of the aspects we have discussed, at least briefly, is the general economy of the different options. As we have mentioned the M&A has advantages for both parties but is less beneficial to the economy in general than the maturity of the TSU into a complete firm by itself.

Having discussed the economic benefits, we have moved to discuss the government support tools. Before we discussed the assistance given to the specific move, which is our topic, we had a lengthy discussion regarding the types and aspects of government support in general, and although regarding specific assistance. We discuss the pros and cons of government support and the specific role of support programs.

We continue the discussion by presenting several tools that are in use.

The following chapter, chapter 7 is a case study. In the case study we present the story of SeeTree Ltd., a small TSU making the move towards a complete company, but not in the usual way, presenting another alternative. The model of ST, in which the move is performed along the life of the TSU and not as a separate move. The result of the model is a smoother transition. The model ends the discussion, before we move to this summary.

There is no preferred model. The creation of an ecosystem that allows the establishment of many successful TSUs and hopefully the success of enough of them in the long run to create a stable and growing economy, depends among others on the motivation of founders and entrepreneurs, in which the benefits from a potential M&A play a big role. Therefore, although there are some mitigating influences as the processes pushes towards less small and medium companies, on the other hands, it does create a push for the creation in new TSUs.

The maturation of the TSU to a complete firm, has its own benefits, especially in the creation of a large number of small firms. In the relatively bigger demand for HR in this process compared to the M&A option. It is this option which also allows the creation of experienced cadre of managers along the entire maturation process of the start-ups. This is an asset by itself for every ecosystem.

Glossary

Advisory Board (AD): The AD is an organ of the firm that can be used to advise the manager on different issues where decisions or the ability to reach new contacts are required.

Beta-Site Testing: This is an advanced prototype testing that is carried out at a potential client site, or otherwise mimicking real market conditions for testing the product and its value to the clients.

Bootstrap: A new technology-based enterprise with economic gain intentions basing its resources on economic gains from a commercial activity while endeavoring to perform R&D as the basis for future growth.

Complete Firm: A company for profit, deriving its resources mainly (not exclusively) from its daily operation as revenues from the business it is conducting. Deriving the revenues from the business requires a complex collection of different activities and this requires in turn complex set of managerial capabilities.

Ecosystem: The economic, social, and regulatory environment that the TSU come into interactions with. This includes funding sources, employment, and social laws, as well as other relevant regulatory measures pertaining to the activity and livelihood of the TSU.

Environmental Fit: By defining an enterprise or any organization, a definition of the environment in which the organization operates is formed. Borrowing from biological evolution, the level of adaptation to the environment of the organization is the environmental fit of the organization. The higher the fit, the more adapt the organization to the demands and opportunities of the environment in which it operates.

Financial Aid: Support either by grant, loan or loan guarantee of funds aimed at assisting (in our case) the TSU in overcoming barriers that are not well supported by private market support.

Firm Growth: The increase in the assets (tangible and non-tangible) of the firm. It can be expressed in several ways, increase in income, sales, profits, market share etc., or in the number of employees investment in infrastructure (sites, buildings, and machines), as well as in the intellectual property and reputation.

Israeli Innovation Authority: A corporate created by law in Israel in 2016 taking over the functions of several units within the government, and without, supporting innovation actions. It replaces the Office of the Chief Scientist of the Ministry of economic development and is operating the programs formerly operated by that office.

Market Entry: The introduction into the market of a new product or technology. This is the first time the specific product or technology (of the specific enterprise) is presented to the market (market segment) and this would determine in most cases how the product will be incorporated by the market, and the market share it will acquire finally. This is a strategic action and requires both planning and market understanding as well as risk management.

Market Failure: A lack of positive action from the private market, or private entities to support a commercial activity. This is often due to perceived high risk, or lack of clear benefits resulting to the private entities involved. It means that certain activities are not supported by private entities even if these activities would benefit the economy in general. In that case it is often up-to public (government) entities to get involved and make supporting the activities more attractive to the private sector or support the activities directly. This is done in order to benefit the economy in general.

Merger and Acquisition (M&A): The incorporation of one enterprise into another, through ownership but mainly through joint management. While in most cases it would be the bigger company that acquires the TSU, for our discussion it is not the size that is important, but rather that the Complete Firm takes managerial control over the TSU.

Non-Financial Aid: A similarly aimed support which is not financial but can take other forms (e.g., mentoring support, general lenient for TSU regulation).

Non-Technical SU: A business-oriented SU without recourse to special or proprietary technology, (e.g., a restaurant or a shop).

Organizational Culture: For the purpose of our discussion here. The organizational culture shall refer to the forms of the interactions between the enterprise organs and the collection of customs and values of the enterprise.

Organizational Structure: For the purposes of our discussion here the organizational structure of an enterprise, refers to the organs in the enterprise and their interactions.

Perpetual/Repetitive Entrepreneur: An entrepreneur who like the entrepreneurial stage so much that each time the TSU matures, the entrepreneur leaves the enterprise in favor of starting another one.

Social Start-Up: A non-for-profit SU intended to forward some social goal, raising funds in order to finance its activities as these do not create the resources it needs for further sustenance.

Start-Up (SU): A general reference to a new enterprise whether for profit or not. It is mostly used for new companies or businesses formed and referrers to the early stages of their lives.

Technological Start-Up (TSU): A special case of the SU characterized by being based mostly on technology and preoccupied by performing mainly technological R&D and being economically supported mainly by investment and not by income generated from business activity. In may researches and publications it is this type that is referred to as Start-Up. For our discussion it is important to differentiate between a TSU and a general SU which can be a social or business-oriented entity not dealing with technology or R&D at all.

Technology Innovation Incubators Program: A national program supporting TSUs in technology incubators. The incubators are firms selected by tender for a certain period in which they are selecting TSUs and together applying for financial support for the TSUs from the IIA. The incubators, receive for their services, equity, as well as additional equity for supplementary financial support. This program, started in 1990 has undergone several changes and is considered an engine for the creation of TSUs.

Transient Capabilities: The capabilities that are linked directly to specific personnel, based on their experience or training, and which exists in the firm as long as the specific person is employed. When the personnel change, the capabilities will change with the people leaving and the new people replacing them. These capabilities are not reflected in the processes and customs of the firm which remain in force even with personnel change.

Value Creation Process: For the purpose of our discussion this is the process in which by advancing along the steps of the development chain from basic technology to finished product in the market, a TSU increases its value for investment.

Compilation of References

Ács, Z. J., Autio, E., & Szerb, L. (2014). National Systems of Entrepreneurship: Measurement Issues and Policy Implications. *Research Policy, 43*(3), 476–494. doi:10.1016/j.respol.2013.08.016

Almeida, F., & Miguel-Oliveira, J. (2022). The Role of Intrapreneurship in Portuguese Startups. *Periodica Polytechnica Social and Management Sciences, 30*(1), 70–79. doi:10.3311/PPso.17796

Asheim, B. T., Smith, H. L., & Oughton, C. (2011). Regional Innovation Systems: Theory, Empirics and Policy. *Regional Studies, 45*(7), 875–891. doi:10.1080/00343404.2011.596701

Audretsch, D. B., & Belitski, M. (2016). Entrepreneurial Ecosystems in Cities: Establishing the Framework Conditions. *The Journal of Technology Transfer, 42*(5), 1030–1051. doi:10.100710961-016-9473-8

Audretsch, D., & Keilbach, M. (2004). Entrepreneurship Capital and Economic Performance. *Regional Studies, 38*(8), 949–959. doi:10.1080/0034340042000280956

Baltrunaite, V., & Sekliuckiene, J. (2020). The Use of Organizational Learning Practices in Start-Ups growth. *Entrepreneurial Business and Economics Review, 8*(1), 71–89. doi:10.15678/EBER.2020.080104

Barr, P. S., Stimpert, J. L., & Huff, A. S. (1992). Cognitive change, strategic action and organizational renewal. *Strategic Management Journal, 13*(Special Issue), 15–36. doi:10.1002mj.4250131004

Baum, J. R., Locke, E. A., & Smith, K. G. (2001). A Multidimensional Model of Venture Growth. *Academy of Management Journal, 44*(2), 292–303. doi:10.2307/3069456

Brown, S. I., & Eisenhardt, K. M. (1998). *Competing on the Edge*. Harvard Business School Press.

Brunkhorst, H. (2020). "Everything is now in order" – Groupthink, ideology and practical-critical activity in the European financial crisis and beyond. *Journal of European Public Policy, 27*(9), 1406–1423. doi:10.1080/13501763.2020.1773515

Calantone, R. J., Cavusgil, S. T., & Zhao, Y. (2002). Learning Orientation, Firm Innovation Capability, and Firm Performance. *Industrial Marketing Management, 31*(6), 515–524. doi:10.1016/S0019-8501(01)00203-6

Cohen, B. (2006). Sustainable Valley Entrepreneurial Ecosystems. *Business Strategy and the Environment*, *15*(1), 1–14. doi:10.1002/bse.428

Delmar, F., & Shane, S. (2006). Does Experience Matter? The Effect of Founding Team Experience on the Survival and Sales of Newly Founded Ventures. *Strategic Organization*, *4*(3), 215–247. doi:10.1177/1476127006066596

Denison, D. R. (1990). Corporate Culture and Organizational Effectiveness. New York: John Wiley & Sons.

Denison, D. R., Janovics, J., & Young, J. (2005). *Diagnosing Organizational Cultures: Validating a Model and Method-Working Paper*. Denison Consulting.

Entrialgo, M., & Iglesias, V. (2017). Are the Intentions to Entrepreneurship of Men and Women Shaped Differently? The Impact of Entrepreneurial Role-Model Exposure and Entrepreneurship Education. *Entrepreneurship Research Journal*, *20170013*(1), 1–14. doi:10.1515/erj-2017-0013

EU. (2000). *Lisbon European Council 23 and 24 March 2000*. https://www.europarl.europa.eu/summits/lis1_en.htm#1

EU. (2003). *Communication from the Commission; Investing in research: An action plan for Europe*. COM (2003) 226 final. Available online at: https://ec.europa.eu/invest-in-research/pdf/226/en.pdf

EU. (2003). *Communication from the Commission; Investing in research: An action plan for Europe*. COM (2003) 226 final. https://ec.europa.eu/invest-in-research/pdf/226/en.pdf

Feld, B. (2012). *Startup Communities: Building an Entrepreneurial Ecosystem in Your City*. Wiley. doi:10.1002/9781119204459

Folz, J. M. (2003). Managing two brands for success: Peugeot and Citroën. In B. Gottschalk & R. G. Kalmbach (Eds.), *Markenmanagement in der Automobilindustrie*. Gabler Verlag. doi:10.1007/978-3-322-92081-2_16

Gabai, U., Ehrlich, E., Abdel Razek, K., & Krieger, N. (2020). *High-Tech Human Capital Report 2019*. Israel Innovation Authority and Start-Up Nation Cantral.

Gompers, P. A. (1995). Optimal Investment, Monitoring, and the Staging of Venture Capital. *The Journal of Finance*, *50*(5), 1461–1489. doi:10.1111/j.1540-6261.1995.tb05185.x

Groupthink, B. (2013). Collective Delusions in Organizations and Markets. *The Review of Economic Studies*, *80*(2(283)), 429–462.

Hasliza, A. (2014). The Growth of Innovative Performance among SMEs: Leveraging on Organisational Culture and Innovative Human Capital. *Journal of Small Business and Entrepreneurship Development*, *2*(1), 107–125.

Hughes, M., & Morgan, R. E. (2007). Deconstructing the Relationship between Entrepreneurial Orientation and Business Performance at the Embryonic Stage of Firm Growth. *Industrial Marketing Management*, *36*(5), 651–661. doi:10.1016/j.indmarman.2006.04.003

Israeli Innovation Authority. (2022). *Technological Innovation incubators program*. Downloaded on 19 December 2022 from https://innovationisrael.org.il/en/program/incubators-incentive-program

Jaskyte, K. (2004). Transformational Leadership, Organizational Culture, and Innovativeness in Nonprofit Organizations. *Nonprofit Management & Leadership*, *15*(2), 153–168. doi:10.1002/nml.59

Jiménez, D. J., & Sanz-Valle, R. (2011). Innovation, Organizational Learning, and Performance. *Journal of Business Research*, *64*(4), 408–417. doi:10.1016/j.jbusres.2010.09.010

Kang, H. D. (2018). A Start-Up's R&D Stages and the Evolution of Financing Sources: Evidence from the Biotechnology Industry. *Entrepreneurship Research Journal*, *20170159*(3), 1–19. doi:10.1515/erj-2017-0159

Korbet, R., & Analyst, M. (2019). *The State of the Israeli Ecosystem in 2018*. Start-Up Nation Central.

Lerner, J. (2000). The government as venture capitalist: The long-run impact of the SBIR program. *Journal of Private Equity*, *3*(2), 55–78. doi:10.3905/jpe.2000.319960

Luukkonen, T. (1988). The difficulties in assessing the impact of EU Framework Programs. *Research Policy*, *27*(6), 599–610. doi:10.1016/S0048-7333(98)00058-4

Luukkonen, T. (2000). Additionality of EU Framework Programs. *Research Policy*, *29*(6), 711–724. doi:10.1016/S0048-7333(99)00041-4

Mustafa, M., Lundmark, E., & Ramos, H. M. (2016). Untangling the Relationship between Human Resource Management and Corporate Entrepreneurship: The Mediating Effect of Middle Managers' Knowledge Sharing. *Entrepreneurship Research Journal*, *6*(3), 273–295. doi:10.1515/erj-2015-0004

Nason, R. S., & Wiklund, J. (2018). An Assessment of Resource-Based Theorizing on Firm Growth and Suggestions for the Future. *Journal of Management*, *44*(1), 32–60. doi:10.1177/0149206315610635

Noy, E., & Ellis, S. (2003a). Risk: A neglected component of strategy formulation. *Journal of Managerial Psychology*, *18*(7), 691–707. doi:10.1108/02683940310502395

Noy, E., & Ellis, S. (2003b). Corporate Risk Strategy: Does it Vary Across Business Activities? *European Management Journal*, *21*(1), 119–128. doi:10.1016/S0263-2373(02)00159-7

Owens, T., & Fernandez, O. (2014). *The Lean Enterprise: How Corporations Can Innovate Like Start-Ups*. Wiley. doi:10.1002/9781118854006

Penrose, E. T. (1959). *The Theory of the Growth of the Firm*. John Wiley.

Porath, A. (2010). *Collaborative Research in Management: Inside out*. SAGE Publications.

Porath, A. (2017). *Supporting Innovation Through National R&D Programs: Emerging Research and Opportunities*. IGI Global.

Porath, A., Ramos, I., & Rahman, H. (2015). Different approaches to technology transfer by government and academy. *World Review of Entrepreneurship, Management and Sustainable Development*, *11*(2/3), 296–307. doi:10.1504/WREMSD.2015.068585

Porter, M. (2013). *The case for letting business solve social problems*. https://www.ted.com/talks/michael_porter_the_case_for_letting_business_solve_social_problems

Puri, M., & Zarutskie, R. (2012). On the Life Cycle Dynamics of Venture-Capital and Nonventure-Capital Backed Companies. *The Journal of Finance*, *67*(6), 2247–2293. doi:10.1111/j.1540-6261.2012.01786.x

Samila, S., & Sorenson, O. (2011). Venture capital, entrepreneurship, and economic growth. *The Review of Economics and Statistics*, *93*(1), 338–349. doi:10.1162/REST_a_00066

Santi, C., & Neves, G. T. (2020). Startups and prominent business. In A. G. Rodrigues (Ed.), *New and Future Developments in Microbial Biotechnology and Bioengineering* (pp. 209–223). Elsevier. doi:10.1016/B978-0-444-64301-8.00010-X

Sarma, S., & Marszalek, J. (2019). New Venture Growth: Role of Ecosystem Elements and Prior Experience. *Entrepreneurship Research Journal, 10*(1). . doi:10.1515/erj-2018-0215

Scalera, F. A. (2014). Internationalisation Strategies in the Automotive Sector. The PSA Peugeot Citroen and Fiat Auto Groups: A Comparison between the two Models. *JIEB*, (2), 28–37.

Schneider, S. C. (1998). National Vs. Corporate Culture: Implications for Human Resource Management. *Human Resource Management*, *27*(2), 231–246. doi:10.1002/hrm.3930270207

SeeTree Ltd. (2020). *One Pager for Customers.* Author.

Shiller, R. J. (2017). Narrative Economics. *The American Economic Review*, *107*(4), 967–1004. doi:10.1257/aer.107.4.967

Spigel, B., & Harrison, R. (2018). Toward a Process Theory of Entrepreneurial Ecosystems. *Strategic Entrepreneurship Journal*, *12*(1), 151–168. doi:10.1002ej.1268

Storey, D. J. (1991). The Birth of New Firms – Does Unemployment Matter? A Review of the Evidence. *Small Business Economics*, *3*(3), 167–178. doi:10.1007/BF00400022

Taleb, N. N. (2012). Antifragile. Things that gain from disorder. Random House.

Taleb, N. N. (2012). *Antifragile. Things that gain from disorder*. Random House Trade Paperbacks.

Tsegay, B. K., Shi, C. S., & Wu, W. W. (2016). The Impact of Innovation Strategy on Organizational Learning and Innovation Performance: Do Firm Size and Ownership Type Make a Difference? *South African Journal of Industrial Engineering*, *27*(1), 125–136.

Yin, B., & Lou, J. (2018). How Do Accelerators Select Startups? Shifting Decision Criteria Across Stages. *IEEE Transactions on Engineering Management*, *65*(4), 574–589. doi:10.1109/TEM.2018.2791501

Yiu, D., & Lau, C. (2008). Corporate Entrepreneurship as Resource Capital Configuration in Emerging Market Firms. *Entrepreneurship Theory and Practice*, *32*(1), 37–57. doi:10.1111/j.1540-6520.2007.00215.x

Zahra, S. A. (1995). Corporate Entrepreneurship and Financial Performance: The Case of Management Leveraged Buyouts. *Journal of Business Venturing*, *10*(3), 225–247. doi:10.1016/0883-9026(94)00024-O

Zahra, S., Jennings, S., & Kuratko, D. (1999). The Antecedents and Consequences of Firm-Level Entrepreneurship: The State of the Field. *Entrepreneurship Theory and Practice*, *24*(2), 45–66. doi:10.1177/104225879902400205

Zhang, Z., & Jia, M. (2010). Using Social Exchange Theory to Predict the Effects of High-Performance Human Resource Practices on Corporate Entrepreneurship: Evidence from China. *Human Resource Management*, *49*(4), 743–765. doi:10.1002/hrm.20378

Related References

To continue our tradition of advancing information science and technology research, we have compiled a list of recommended IGI Global readings. These references will provide additional information and guidance to further enrich your knowledge and assist you with your own research and future publications.

Abdul Razak, R., & Mansor, N. A. (2021). Instagram Influencers in Social Media-Induced Tourism: Rethinking Tourist Trust Towards Tourism Destination. In M. Dinis, L. Bonixe, S. Lamy, & Z. Breda (Eds.), *Impact of New Media in Tourism* (pp. 135-144). IGI Global. https://doi.org/10.4018/978-1-7998-7095-1.ch009

Abir, T., & Khan, M. Y. (2022). Importance of ICT Advancement and Culture of Adaptation in the Tourism and Hospitality Industry for Developing Countries. In C. Ramos, S. Quinteiro, & A. Gonçalves (Eds.), *ICT as Innovator Between Tourism and Culture* (pp. 30–41). IGI Global. https://doi.org/10.4018/978-1-7998-8165-0.ch003

Abtahi, M. S., Behboudi, L., & Hasanabad, H. M. (2017). Factors Affecting Internet Advertising Adoption in Ad Agencies. *International Journal of Innovation in the Digital Economy*, *8*(4), 18–29. doi:10.4018/IJIDE.2017100102

Afenyo-Agbe, E., & Mensah, I. (2022). Principles, Benefits, and Barriers to Community-Based Tourism: Implications for Management. In I. Mensah & E. Afenyo-Agbe (Eds.), *Prospects and Challenges of Community-Based Tourism and Changing Demographics* (pp. 1–29). IGI Global. doi:10.4018/978-1-7998-7335-8.ch001

Agbo, V. M. (2022). Distributive Justice Issues in Community-Based Tourism. In I. Mensah & E. Afenyo-Agbe (Eds.), *Prospects and Challenges of Community-Based Tourism and Changing Demographics* (pp. 107–129). IGI Global. https://doi.org/10.4018/978-1-7998-7335-8.ch005

Agrawal, S. (2017). The Impact of Emerging Technologies and Social Media on Different Business(es): Marketing and Management. In O. Rishi & A. Sharma (Eds.), *Maximizing Business Performance and Efficiency Through Intelligent Systems* (pp. 37–49). Hershey, PA: IGI Global. doi:10.4018/978-1-5225-2234-8.ch002

Ahmad, A., & Johari, S. (2022). Georgetown as a Gastronomy Tourism Destination: Visitor Awareness Towards Revisit Intention of Nasi Kandar Restaurant. In M. Valeri (Ed.), *New Governance and Management in Touristic Destinations* (pp. 71–83). IGI Global. https://doi.org/10.4018/978-1-6684-3889-3.ch005

Alkhatib, G., & Bayouq, S. T. (2021). A TAM-Based Model of Technological Factors Affecting Use of E-Tourism. *International Journal of Tourism and Hospitality Management in the Digital Age*, *5*(2), 50–67. https://doi.org/10.4018/IJTHMDA.20210701.oa1

Altinay Ozdemir, M. (2021). Virtual Reality (VR) and Augmented Reality (AR) Technologies for Accessibility and Marketing in the Tourism Industry. In C. Eusébio, L. Teixeira, & M. Carneiro (Eds.), *ICT Tools and Applications for Accessible Tourism* (pp. 277-301). IGI Global. https://doi.org/10.4018/978-1-7998-6428-8.ch013

Anantharaman, R. N., Rajeswari, K. S., Angusamy, A., & Kuppusamy, J. (2017). Role of Self-Efficacy and Collective Efficacy as Moderators of Occupational Stress Among Software Development Professionals. *International Journal of Human Capital and Information Technology Professionals*, *8*(2), 45–58. doi:10.4018/IJHCITP.2017040103

Aninze, F., El-Gohary, H., & Hussain, J. (2018). The Role of Microfinance to Empower Women: The Case of Developing Countries. *International Journal of Customer Relationship Marketing and Management*, *9*(1), 54–78. doi:10.4018/IJCRMM.2018010104

Antosova, G., Sabogal-Salamanca, M., & Krizova, E. (2021). Human Capital in Tourism: A Practical Model of Endogenous and Exogenous Territorial Tourism Planning in Bahía Solano, Colombia. In V. Costa, A. Moura, & M. Mira (Eds.), *Handbook of Research on Human Capital and People Management in the Tourism Industry* (pp. 282–302). IGI Global. https://doi.org/10.4018/978-1-7998-4318-4.ch014

Arsenijević, O. M., Orčić, D., & Kastratović, E. (2017). Development of an Optimization Tool for Intangibles in SMEs: A Case Study from Serbia with a Pilot Research in the Prestige by Milka Company. In M. Vemić (Ed.), *Optimal Management Strategies in Small and Medium Enterprises* (pp. 320–347). Hershey, PA: IGI Global. doi:10.4018/978-1-5225-1949-2.ch015

Aryanto, V. D., Wismantoro, Y., & Widyatmoko, K. (2018). Implementing Eco-Innovation by Utilizing the Internet to Enhance Firm's Marketing Performance: Study of Green Batik Small and Medium Enterprises in Indonesia. *International Journal of E-Business Research, 14*(1), 21–36. doi:10.4018/IJEBR.2018010102

Asero, V., & Billi, S. (2022). New Perspective of Networking in the DMO Model. In M. Valeri (Ed.), *New Governance and Management in Touristic Destinations* (pp. 105–118). IGI Global. https://doi.org/10.4018/978-1-6684-3889-3.ch007

Atiku, S. O., & Fields, Z. (2017). Multicultural Orientations for 21st Century Global Leadership. In N. Baporikar (Ed.), *Management Education for Global Leadership* (pp. 28–51). Hershey, PA: IGI Global. doi:10.4018/978-1-5225-1013-0.ch002

Atiku, S. O., & Fields, Z. (2018). Organisational Learning Dimensions and Talent Retention Strategies for the Service Industries. In N. Baporikar (Ed.), *Global Practices in Knowledge Management for Societal and Organizational Development* (pp. 358–381). Hershey, PA: IGI Global. doi:10.4018/978-1-5225-3009-1.ch017

Atsa'am, D. D., & Kuset Bodur, E. (2021). Pattern Mining on How Organizational Tenure Affects the Psychological Capital of Employees Within the Hospitality and Tourism Industry: Linking Employees' Organizational Tenure With PsyCap. *International Journal of Tourism and Hospitality Management in the Digital Age, 5*(2), 17–28. https://doi.org/10.4018/IJTHMDA.2021070102

Ávila, L., & Teixeira, L. (2018). The Main Concepts Behind the Dematerialization of Business Processes. In M. Khosrow-Pour, D.B.A. (Ed.), Encyclopedia of Information Science and Technology, Fourth Edition (pp. 888-898). Hershey, PA: IGI Global. https://doi.org/ doi:10.4018/978-1-5225-2255-3.ch076

Ayorekire, J., Mugizi, F., Obua, J., & Ampaire, G. (2022). Community-Based Tourism and Local People's Perceptions Towards Conservation: The Case of Queen Elizabeth Conservation Area, Uganda. In I. Mensah & E. Afenyo-Agbe (Eds.), *Prospects and Challenges of Community-Based Tourism and Changing Demographics* (pp. 56–82). IGI Global. https://doi.org/10.4018/978-1-7998-7335-8.ch003

Baleiro, R. (2022). Tourist Literature and the Architecture of Travel in Olga Tokarczuk and Patti Smith. In R. Baleiro & R. Pereira (Eds.), *Global Perspectives on Literary Tourism and Film-Induced Tourism* (pp. 202-216). IGI Global. https://doi.org/10.4018/978-1-7998-8262-6.ch011

Barat, S. (2021). Looking at the Future of Medical Tourism in Asia. *International Journal of Tourism and Hospitality Management in the Digital Age*, *5*(1), 19–33. https://doi.org/10.4018/IJTHMDA.2021010102

Barbosa, C. A., Magalhães, M., & Nunes, M. R. (2021). Travel Instagramability: A Way of Choosing a Destination? In M. Dinis, L. Bonixe, S. Lamy, & Z. Breda (Eds.), *Impact of New Media in Tourism* (pp. 173-190). IGI Global. https://doi.org/10.4018/978-1-7998-7095-1.ch011

Bari, M. W., & Khan, Q. (2021). Pakistan as a Destination of Religious Tourism. In E. Alaverdov & M. Bari (Eds.), *Global Development of Religious Tourism* (pp. 1-10). IGI Global. https://doi.org/10.4018/978-1-7998-5792-1.ch001

Bartens, Y., Chunpir, H. I., Schulte, F., & Voß, S. (2017). Business/IT Alignment in Two-Sided Markets: A COBIT 5 Analysis for Media Streaming Business Models. In S. De Haes & W. Van Grembergen (Eds.), *Strategic IT Governance and Alignment in Business Settings* (pp. 82–111). Hershey, PA: IGI Global. doi:10.4018/978-1-5225-0861-8.ch004

Bashayreh, A. M. (2018). Organizational Culture and Organizational Performance. In W. Lee & F. Sabetzadeh (Eds.), *Contemporary Knowledge and Systems Science* (pp. 50–69). Hershey, PA: IGI Global. doi:10.4018/978-1-5225-5655-8.ch003

Bechthold, L., Lude, M., & Prügl, R. (2021). Crisis Favors the Prepared Firm: How Organizational Ambidexterity Relates to Perceptions of Organizational Resilience. In A. Zehrer, G. Glowka, K. Schwaiger, & V. Ranacher-Lackner (Eds.), *Resiliency Models and Addressing Future Risks for Family Firms in the Tourism Industry* (pp. 178–205). IGI Global. https://doi.org/10.4018/978-1-7998-7352-5.ch008

Bedford, D. A. (2018). Sustainable Knowledge Management Strategies: Aligning Business Capabilities and Knowledge Management Goals. In N. Baporikar (Ed.), *Global Practices in Knowledge Management for Societal and Organizational Development* (pp. 46–73). Hershey, PA: IGI Global. doi:10.4018/978-1-5225-3009-1.ch003

Bekjanov, D., & Matyusupov, B. (2021). Influence of Innovative Processes in the Competitiveness of Tourist Destination. In J. Soares (Ed.), *Innovation and Entrepreneurial Opportunities in Community Tourism* (pp. 243–263). IGI Global. https://doi.org/10.4018/978-1-7998-4855-4.ch014

Bharwani, S., & Musunuri, D. (2018). Reflection as a Process From Theory to Practice. In M. Khosrow-Pour, D.B.A. (Ed.), Encyclopedia of Information Science and Technology, Fourth Edition (pp. 1529-1539). Hershey, PA: IGI Global. doi:10.4018/978-1-5225-2255-3.ch132

Bhatt, G. D., Wang, Z., & Rodger, J. A. (2017). Information Systems Capabilities and Their Effects on Competitive Advantages: A Study of Chinese Companies. *Information Resources Management Journal*, *30*(3), 41–57. doi:10.4018/IRMJ.2017070103

Bhushan, M., & Yadav, A. (2017). Concept of Cloud Computing in ESB. In R. Bhadoria, N. Chaudhari, G. Tomar, & S. Singh (Eds.), *Exploring Enterprise Service Bus in the Service-Oriented Architecture Paradigm* (pp. 116–127). Hershey, PA: IGI Global. doi:10.4018/978-1-5225-2157-0.ch008

Bhushan, S. (2017). System Dynamics Base-Model of Humanitarian Supply Chain (HSCM) in Disaster Prone Eco-Communities of India: A Discussion on Simulation and Scenario Results. *International Journal of System Dynamics Applications*, *6*(3), 20–37. doi:10.4018/IJSDA.2017070102

Binder, D., & Miller, J. W. (2021). A Generations' Perspective on Employer Branding in Tourism. In V. Costa, A. Moura, & M. Mira (Eds.), *Handbook of Research on Human Capital and People Management in the Tourism Industry* (pp. 152–174). IGI Global. https://doi.org/10.4018/978-1-7998-4318-4.ch008

Birch Freeman, A. A., Mensah, I., & Antwi, K. B. (2022). Smiling vs. Frowning Faces: Community Participation for Sustainable Tourism in Ghanaian Communities. In I. Mensah & E. Afenyo-Agbe (Eds.), *Prospects and Challenges of Community-Based Tourism and Changing Demographics* (pp. 83–106). IGI Global. https://doi.org/10.4018/978-1-7998-7335-8.ch004

Biswas, A., & De, A. K. (2017). On Development of a Fuzzy Stochastic Programming Model with Its Application to Business Management. In S. Trivedi, S. Dey, A. Kumar, & T. Panda (Eds.), *Handbook of Research on Advanced Data Mining Techniques and Applications for Business Intelligence* (pp. 353–378). Hershey, PA: IGI Global. doi:10.4018/978-1-5225-2031-3.ch021

Boragnio, A., & Faracce Macia, C. (2021). "Taking Care of Yourself at Home": Use of E-Commerce About Food and Care During the COVID-19 Pandemic in the City of Buenos Aires. In M. Korstanje (Ed.), *Socio-Economic Effects and Recovery Efforts for the Rental Industry: Post-COVID-19 Strategies* (pp. 45–71). IGI Global. https://doi.org/10.4018/978-1-7998-7287-0.ch003

Borges, V. D. (2021). Happiness: The Basis for Public Policy in Tourism. In A. Perinotto, V. Mayer, & J. Soares (Eds.), *Rebuilding and Restructuring the Tourism Industry: Infusion of Happiness and Quality of Life* (pp. 1–25). IGI Global. https://doi.org/10.4018/978-1-7998-7239-9.ch001

Bücker, J., & Ernste, K. (2018). Use of Brand Heroes in Strategic Reputation Management: The Case of Bacardi, Adidas, and Daimler. In A. Erdemir (Ed.), *Reputation Management Techniques in Public Relations* (pp. 126–150). Hershey, PA: IGI Global. doi:10.4018/978-1-5225-3619-2.ch007

Buluk Eşitti, B. (2021). COVID-19 and Alternative Tourism: New Destinations and New Tourism Products. In M. Demir, A. Dalgıç, & F. Ergen (Eds.), *Handbook of Research on the Impacts and Implications of COVID-19 on the Tourism Industry* (pp. 786–805). IGI Global. https://doi.org/10.4018/978-1-7998-8231-2.ch038

Bureš, V. (2018). Industry 4.0 From the Systems Engineering Perspective: Alternative Holistic Framework Development. In R. Brunet-Thornton & F. Martinez (Eds.), *Analyzing the Impacts of Industry 4.0 in Modern Business Environments* (pp. 199–223). Hershey, PA: IGI Global. doi:10.4018/978-1-5225-3468-6.ch011

Buzady, Z. (2017). Resolving the Magic Cube of Effective Case Teaching: Benchmarking Case Teaching Practices in Emerging Markets – Insights from the Central European University Business School, Hungary. In D. Latusek (Ed.), *Case Studies as a Teaching Tool in Management Education* (pp. 79–103). Hershey, PA: IGI Global. doi:10.4018/978-1-5225-0770-3.ch005

Camillo, A. (2021). *Legal Matters, Risk Management, and Risk Prevention: From Forming a Business to Legal Representation.* IGI Global. doi:10.4018/978-1-7998-4342-9.ch004

Căpusneanu, S., & Topor, D. I. (2018). Business Ethics and Cost Management in SMEs: Theories of Business Ethics and Cost Management Ethos. In I. Oncioiu (Ed.), *Ethics and Decision-Making for Sustainable Business Practices* (pp. 109–127). Hershey, PA: IGI Global. doi:10.4018/978-1-5225-3773-1.ch007

Chan, R. L., Mo, P. L., & Moon, K. K. (2018). Strategic and Tactical Measures in Managing Enterprise Risks: A Study of the Textile and Apparel Industry. In K. Strang, M. Korstanje, & N. Vajjhala (Eds.), *Research, Practices, and Innovations in Global Risk and Contingency Management* (pp. 1–19). Hershey, PA: IGI Global. doi:10.4018/978-1-5225-4754-9.ch001

Charlier, S. D., Burke-Smalley, L. A., & Fisher, S. L. (2018). Undergraduate Programs in the U.S: A Contextual and Content-Based Analysis. In J. Mendy (Ed.), *Teaching Human Resources and Organizational Behavior at the College Level* (pp. 26–57). Hershey, PA: IGI Global. doi:10.4018/978-1-5225-2820-3.ch002

Chumillas, J., Güell, M., & Quer, P. (2022). The Use of ICT in Tourist and Educational Literary Routes: The Role of the Guide. In C. Ramos, S. Quinteiro, & A. Gonçalves (Eds.), *ICT as Innovator Between Tourism and Culture* (pp. 15–29). IGI Global. https://doi.org/10.4018/978-1-7998-8165-0.ch002

Dahlberg, T., Kivijärvi, H., & Saarinen, T. (2017). IT Investment Consistency and Other Factors Influencing the Success of IT Performance. In S. De Haes & W. Van Grembergen (Eds.), *Strategic IT Governance and Alignment in Business Settings* (pp. 176–208). Hershey, PA: IGI Global. doi:10.4018/978-1-5225-0861-8.ch007

Damnjanović, A. M. (2017). Knowledge Management Optimization through IT and E-Business Utilization: A Qualitative Study on Serbian SMEs. In M. Vemić (Ed.), *Optimal Management Strategies in Small and Medium Enterprises* (pp. 249–267). Hershey, PA: IGI Global. doi:10.4018/978-1-5225-1949-2.ch012

Daneshpour, H. (2017). Integrating Sustainable Development into Project Portfolio Management through Application of Open Innovation. In M. Vemić (Ed.), *Optimal Management Strategies in Small and Medium Enterprises* (pp. 370–387). Hershey, PA: IGI Global. doi:10.4018/978-1-5225-1949-2.ch017

Daniel, A. D., & Reis de Castro, V. (2018). Entrepreneurship Education: How to Measure the Impact on Nascent Entrepreneurs. In A. Carrizo Moreira, J. Guilherme Leitão Dantas, & F. Manuel Valente (Eds.), *Nascent Entrepreneurship and Successful New Venture Creation* (pp. 85–110). Hershey, PA: IGI Global. doi:10.4018/978-1-5225-2936-1.ch004

David, R., Swami, B. N., & Tangirala, S. (2018). Ethics Impact on Knowledge Management in Organizational Development: A Case Study. In N. Baporikar (Ed.), *Global Practices in Knowledge Management for Societal and Organizational Development* (pp. 19–45). Hershey, PA: IGI Global. doi:10.4018/978-1-5225-3009-1.ch002

De Uña-Álvarez, E., & Villarino-Pérez, M. (2022). Fostering Ecocultural Resources, Identity, and Tourism in Inland Territories (Galicia, NW Spain). In G. Fernandes (Ed.), *Challenges and New Opportunities for Tourism in Inland Territories: Ecocultural Resources and Sustainable Initiatives* (pp. 1-16). IGI Global. https://doi.org/10.4018/978-1-7998-7339-6.ch001

Delias, P., & Lakiotaki, K. (2018). Discovering Process Horizontal Boundaries to Facilitate Process Comprehension. *International Journal of Operations Research and Information Systems*, *9*(2), 1–31. doi:10.4018/IJORIS.2018040101

Denholm, J., & Lee-Davies, L. (2018). Success Factors for Games in Business and Project Management. In *Enhancing Education and Training Initiatives Through Serious Games* (pp. 34–68). Hershey, PA: IGI Global. doi:10.4018/978-1-5225-3689-5.ch002

Deshpande, M. (2017). Best Practices in Management Institutions for Global Leadership: Policy Aspects. In N. Baporikar (Ed.), *Management Education for Global Leadership* (pp. 1–27). Hershey, PA: IGI Global. doi:10.4018/978-1-5225-1013-0.ch001

Deshpande, M. (2018). Policy Perspectives for SMEs Knowledge Management. In N. Baporikar (Ed.), *Knowledge Integration Strategies for Entrepreneurship and Sustainability* (pp. 23–46). Hershey, PA: IGI Global. doi:10.4018/978-1-5225-5115-7.ch002

Dezdar, S. (2017). ERP Implementation Projects in Asian Countries: A Comparative Study on Iran and China. *International Journal of Information Technology Project Management*, *8*(3), 52–68. doi:10.4018/IJITPM.2017070104

Domingos, D., Respício, A., & Martinho, R. (2017). Reliability of IoT-Aware BPMN Healthcare Processes. In C. Reis & M. Maximiano (Eds.), *Internet of Things and Advanced Application in Healthcare* (pp. 214–248). Hershey, PA: IGI Global. doi:10.4018/978-1-5225-1820-4.ch008

Dosumu, O., Hussain, J., & El-Gohary, H. (2017). An Exploratory Study of the Impact of Government Policies on the Development of Small and Medium Enterprises in Developing Countries: The Case of Nigeria. *International Journal of Customer Relationship Marketing and Management*, *8*(4), 51–62. doi:10.4018/IJCRMM.2017100104

Durst, S., Bruns, G., & Edvardsson, I. R. (2017). Retaining Knowledge in Smaller Building and Construction Firms. *International Journal of Knowledge and Systems Science*, *8*(3), 1–12. doi:10.4018/IJKSS.2017070101

Edvardsson, I. R., & Durst, S. (2017). Outsourcing, Knowledge, and Learning: A Critical Review. *International Journal of Knowledge-Based Organizations*, *7*(2), 13–26. doi:10.4018/IJKBO.2017040102

Edwards, J. S. (2018). Integrating Knowledge Management and Business Processes. In M. Khosrow-Pour, D.B.A. (Ed.), Encyclopedia of Information Science and Technology, Fourth Edition (pp. 5046-5055). Hershey, PA: IGI Global. doi:10.4018/978-1-5225-2255-3.ch437

Eichelberger, S., & Peters, M. (2021). Family Firm Management in Turbulent Times: Opportunities for Responsible Tourism. In A. Zehrer, G. Glowka, K. Schwaiger, & V. Ranacher-Lackner (Eds.), *Resiliency Models and Addressing Future Risks for Family Firms in the Tourism Industry* (pp. 103–124). IGI Global. https://doi.org/10.4018/978-1-7998-7352-5.ch005

Eide, D., Hjalager, A., & Hansen, M. (2022). Innovative Certifications in Adventure Tourism: Attributes and Diffusion. In R. Augusto Costa, F. Brandão, Z. Breda, & C. Costa (Eds.), *Planning and Managing the Experience Economy in Tourism* (pp. 161-175). IGI Global. https://doi.org/10.4018/978-1-7998-8775-1.ch009

Ejiogu, A. O. (2018). Economics of Farm Management. In *Agricultural Finance and Opportunities for Investment and Expansion* (pp. 56–72). Hershey, PA: IGI Global. doi:10.4018/978-1-5225-3059-6.ch003

Ekanem, I., & Abiade, G. E. (2018). Factors Influencing the Use of E-Commerce by Small Enterprises in Nigeria. *International Journal of ICT Research in Africa and the Middle East, 7*(1), 37–53. doi:10.4018/IJICTRAME.2018010103

Ekanem, I., & Alrossais, L. A. (2017). Succession Challenges Facing Family Businesses in Saudi Arabia. In P. Zgheib (Ed.), *Entrepreneurship and Business Innovation in the Middle East* (pp. 122–146). Hershey, PA: IGI Global. doi:10.4018/978-1-5225-2066-5.ch007

El Faquih, L., & Fredj, M. (2017). Ontology-Based Framework for Quality in Configurable Process Models. *Journal of Electronic Commerce in Organizations, 15*(2), 48–60. doi:10.4018/JECO.2017040104

Faisal, M. N., & Talib, F. (2017). Building Ambidextrous Supply Chains in SMEs: How to Tackle the Barriers? *International Journal of Information Systems and Supply Chain Management, 10*(4), 80–100. doi:10.4018/IJISSCM.2017100105

Fernandes, T. M., Gomes, J., & Romão, M. (2017). Investments in E-Government: A Benefit Management Case Study. *International Journal of Electronic Government Research, 13*(3), 1–17. doi:10.4018/IJEGR.2017070101

Figueira, L. M., Honrado, G. R., & Dionísio, M. S. (2021). Human Capital Management in the Tourism Industry in Portugal. In V. Costa, A. Moura, & M. Mira (Eds.), *Handbook of Research on Human Capital and People Management in the Tourism Industry* (pp. 1–19). IGI Global. doi:10.4018/978-1-7998-4318-4.ch001

Gao, S. S., Oreal, S., & Zhang, J. (2018). Contemporary Financial Risk Management Perceptions and Practices of Small-Sized Chinese Businesses. In I. Management Association (Ed.), Global Business Expansion: Concepts, Methodologies, Tools, and Applications (pp. 917-931). Hershey, PA: IGI Global. doi:10.4018/978-1-5225-5481-3.ch041

Garg, R., & Berning, S. C. (2017). Indigenous Chinese Management Philosophies: Key Concepts and Relevance for Modern Chinese Firms. In B. Christiansen & G. Koc (Eds.), *Transcontinental Strategies for Industrial Development and Economic Growth* (pp. 43–57). Hershey, PA: IGI Global. doi:10.4018/978-1-5225-2160-0.ch003

Gencer, Y. G. (2017). Supply Chain Management in Retailing Business. In U. Akkucuk (Ed.), *Ethics and Sustainability in Global Supply Chain Management* (pp. 197–210). Hershey, PA: IGI Global. doi:10.4018/978-1-5225-2036-8.ch011

Gera, R., Arora, S., & Malik, S. (2021). Emotional Labor in the Tourism Industry: Strategies, Antecedents, and Outcomes. In V. Costa, A. Moura, & M. Mira (Eds.), *Handbook of Research on Human Capital and People Management in the Tourism Industry* (pp. 73–91). IGI Global. https://doi.org/10.4018/978-1-7998-4318-4.ch004

Giacosa, E. (2018). The Increasing of the Regional Development Thanks to the Luxury Business Innovation. In L. Carvalho (Ed.), *Handbook of Research on Entrepreneurial Ecosystems and Social Dynamics in a Globalized World* (pp. 260–273). Hershey, PA: IGI Global. doi:10.4018/978-1-5225-3525-6.ch011

Glowka, G., Tusch, M., & Zehrer, A. (2021). The Risk Perception of Family Business Owner-Manager in the Tourism Industry: A Qualitative Comparison of the Intra-Firm Senior and Junior Generation. In A. Zehrer, G. Glowka, K. Schwaiger, & V. Ranacher-Lackner (Eds.), *Resiliency Models and Addressing Future Risks for Family Firms in the Tourism Industry* (pp. 126–153). IGI Global. https://doi.org/10.4018/978-1-7998-7352-5.ch006

Glykas, M., & George, J. (2017). Quality and Process Management Systems in the UAE Maritime Industry. *International Journal of Productivity Management and Assessment Technologies, 5*(1), 20–39. doi:10.4018/IJPMAT.2017010102

Glykas, M., Valiris, G., Kokkinaki, A., & Koutsoukou, Z. (2018). Banking Business Process Management Implementation. *International Journal of Productivity Management and Assessment Technologies, 6*(1), 50–69. doi:10.4018/IJPMAT.2018010104

Gomes, J., & Romão, M. (2017). The Balanced Scorecard: Keeping Updated and Aligned with Today's Business Trends. *International Journal of Productivity Management and Assessment Technologies, 5*(2), 1–15. doi:10.4018/IJPMAT.2017070101

Gomes, J., & Romão, M. (2017). Aligning Information Systems and Technology with Benefit Management and Balanced Scorecard. In S. De Haes & W. Van Grembergen (Eds.), *Strategic IT Governance and Alignment in Business Settings* (pp. 112–131). Hershey, PA: IGI Global. doi:10.4018/978-1-5225-0861-8.ch005

Goyal, A. (2021). Communicating and Building Destination Brands With New Media. In M. Dinis, L. Bonixe, S. Lamy, & Z. Breda (Eds.), *Impact of New Media in Tourism* (pp. 1-20). IGI Global. https://doi.org/10.4018/978-1-7998-7095-1.ch001

Grefen, P., & Turetken, O. (2017). Advanced Business Process Management in Networked E-Business Scenarios. *International Journal of E-Business Research, 13*(4), 70–104. doi:10.4018/IJEBR.2017100105

Guasca, M., Van Broeck, A. M., & Vanneste, D. (2021). Tourism and the Social Reintegration of Colombian Ex-Combatants. In J. da Silva, Z. Breda, & F. Carbone (Eds.), *Role and Impact of Tourism in Peacebuilding and Conflict Transformation* (pp. 66-86). IGI Global. https://doi.org/10.4018/978-1-7998-5053-3.ch005

Haider, A., & Saetang, S. (2017). Strategic IT Alignment in Service Sector. In S. Rozenes & Y. Cohen (Eds.), *Handbook of Research on Strategic Alliances and Value Co-Creation in the Service Industry* (pp. 231–258). Hershey, PA: IGI Global. doi:10.4018/978-1-5225-2084-9.ch012

Hajilari, A. B., Ghadaksaz, M., & Fasghandis, G. S. (2017). Assessing Organizational Readiness for Implementing ERP System Using Fuzzy Expert System Approach. *International Journal of Enterprise Information Systems*, *13*(1), 67–85. doi:10.4018/IJEIS.2017010105

Haldorai, A., Ramu, A., & Murugan, S. (2018). Social Aware Cognitive Radio Networks: Effectiveness of Social Networks as a Strategic Tool for Organizational Business Management. In H. Bansal, G. Shrivastava, G. Nguyen, & L. Stanciu (Eds.), *Social Network Analytics for Contemporary Business Organizations* (pp. 188–202). Hershey, PA: IGI Global. doi:10.4018/978-1-5225-5097-6.ch010

Hall, O. P. Jr. (2017). Social Media Driven Management Education. *International Journal of Knowledge-Based Organizations*, *7*(2), 43–59. doi:10.4018/IJKBO.2017040104

Hanifah, H., Halim, H. A., Ahmad, N. H., & Vafaei-Zadeh, A. (2017). Innovation Culture as a Mediator Between Specific Human Capital and Innovation Performance Among Bumiputera SMEs in Malaysia. In N. Ahmad, T. Ramayah, H. Halim, & S. Rahman (Eds.), *Handbook of Research on Small and Medium Enterprises in Developing Countries* (pp. 261–279). Hershey, PA: IGI Global. doi:10.4018/978-1-5225-2165-5.ch012

Hartlieb, S., & Silvius, G. (2017). Handling Uncertainty in Project Management and Business Development: Similarities and Differences. In Y. Raydugin (Ed.), *Handbook of Research on Leveraging Risk and Uncertainties for Effective Project Management* (pp. 337–362). Hershey, PA: IGI Global. doi:10.4018/978-1-5225-1790-0.ch016

Hass, K. B. (2017). Living on the Edge: Managing Project Complexity. In Y. Raydugin (Ed.), *Handbook of Research on Leveraging Risk and Uncertainties for Effective Project Management* (pp. 177–201). Hershey, PA: IGI Global. doi:10.4018/978-1-5225-1790-0.ch009

Hawking, P., & Carmine Sellitto, C. (2017). Developing an Effective Strategy for Organizational Business Intelligence. In M. Tavana (Ed.), *Enterprise Information Systems and the Digitalization of Business Functions* (pp. 222–237). Hershey, PA: IGI Global. doi:10.4018/978-1-5225-2382-6.ch010

Hawking, P., & Sellitto, C. (2017). A Fast-Moving Consumer Goods Company and Business Intelligence Strategy Development. *International Journal of Enterprise Information Systems*, *13*(2), 22–33. doi:10.4018/IJEIS.2017040102

Hawking, P., & Sellitto, C. (2017). Business Intelligence Strategy: Two Case Studies. *International Journal of Business Intelligence Research*, *8*(2), 17–30. doi:10.4018/IJBIR.2017070102

Hee, W. J., Jalleh, G., Lai, H., & Lin, C. (2017). E-Commerce and IT Projects: Evaluation and Management Issues in Australian and Taiwanese Hospitals. *International Journal of Public Health Management and Ethics*, *2*(1), 69–90. doi:10.4018/IJPHME.2017010104

Hernandez, A. A. (2018). Exploring the Factors to Green IT Adoption of SMEs in the Philippines. *Journal of Cases on Information Technology*, *20*(2), 49–66. doi:10.4018/JCIT.2018040104

Hollman, A., Bickford, S., & Hollman, T. (2017). Cyber InSecurity: A Post-Mortem Attempt to Assess Cyber Problems from IT and Business Management Perspectives. *Journal of Cases on Information Technology*, *19*(3), 42–70. doi:10.4018/JCIT.2017070104

Ibrahim, F., & Zainin, N. M. (2021). Exploring the Technological Impacts: The Case of Museums in Brunei Darussalam. *International Journal of Tourism and Hospitality Management in the Digital Age*, *5*(1), 1–18. https://doi.org/10.4018/IJTHMDA.2021010101

Igbinakhase, I. (2017). Responsible and Sustainable Management Practices in Developing and Developed Business Environments. In Z. Fields (Ed.), *Collective Creativity for Responsible and Sustainable Business Practice* (pp. 180–207). Hershey, PA: IGI Global. doi:10.4018/978-1-5225-1823-5.ch010

Iwata, J. J., & Hoskins, R. G. (2017). Managing Indigenous Knowledge in Tanzania: A Business Perspective. In P. Jain & N. Mnjama (Eds.), *Managing Knowledge Resources and Records in Modern Organizations* (pp. 198–214). Hershey, PA: IGI Global. doi:10.4018/978-1-5225-1965-2.ch012

Jain, P. (2017). Ethical and Legal Issues in Knowledge Management Life-Cycle in Business. In P. Jain & N. Mnjama (Eds.), *Managing Knowledge Resources and Records in Modern Organizations* (pp. 82–101). Hershey, PA: IGI Global. doi:10.4018/978-1-5225-1965-2.ch006

James, S., & Hauli, E. (2017). Holistic Management Education at Tanzanian Rural Development Planning Institute. In N. Baporikar (Ed.), *Management Education for Global Leadership* (pp. 112–136). Hershey, PA: IGI Global. doi:10.4018/978-1-5225-1013-0.ch006

Janošková, M., Csikósová, A., & Čulková, K. (2018). Measurement of Company Performance as Part of Its Strategic Management. In R. Leon (Ed.), *Managerial Strategies for Business Sustainability During Turbulent Times* (pp. 309–335). Hershey, PA: IGI Global. doi:10.4018/978-1-5225-2716-9.ch017

Jean-Vasile, A., & Alecu, A. (2017). Theoretical and Practical Approaches in Understanding the Influences of Cost-Productivity-Profit Trinomial in Contemporary Enterprises. In A. Jean Vasile & D. Nicolò (Eds.), *Sustainable Entrepreneurship and Investments in the Green Economy* (pp. 28–62). Hershey, PA: IGI Global. doi:10.4018/978-1-5225-2075-7.ch002

Joia, L. A., & Correia, J. C. (2018). CIO Competencies From the IT Professional Perspective: Insights From Brazil. *Journal of Global Information Management, 26*(2), 74–103. doi:10.4018/JGIM.2018040104

Juma, A., & Mzera, N. (2017). Knowledge Management and Records Management and Competitive Advantage in Business. In P. Jain & N. Mnjama (Eds.), *Managing Knowledge Resources and Records in Modern Organizations* (pp. 15–28). Hershey, PA: IGI Global. doi:10.4018/978-1-5225-1965-2.ch002

K., I., & A, V. (2018). Monitoring and Auditing in the Cloud. In K. Munir (Ed.), *Cloud Computing Technologies for Green Enterprises* (pp. 318-350). Hershey, PA: IGI Global. https://doi.org/ doi:10.4018/978-1-5225-3038-1.ch013

Kabra, G., Ghosh, V., & Ramesh, A. (2018). Enterprise Integrated Business Process Management and Business Intelligence Framework for Business Process Sustainability. In A. Paul, D. Bhattacharyya, & S. Anand (Eds.), *Green Initiatives for Business Sustainability and Value Creation* (pp. 228–238). Hershey, PA: IGI Global. doi:10.4018/978-1-5225-2662-9.ch010

Kaoud, M. (2017). Investigation of Customer Knowledge Management: A Case Study Research. *International Journal of Service Science, Management, Engineering, and Technology, 8*(2), 12–22. doi:10.4018/IJSSMET.2017040102

Katuu, S. (2018). A Comparative Assessment of Enterprise Content Management Maturity Models. In N. Gwangwava & M. Mutingi (Eds.), *E-Manufacturing and E-Service Strategies in Contemporary Organizations* (pp. 93–118). Hershey, PA: IGI Global. doi:10.4018/978-1-5225-3628-4.ch005

Khan, M. Y., & Abir, T. (2022). The Role of Social Media Marketing in the Tourism and Hospitality Industry: A Conceptual Study on Bangladesh. In C. Ramos, S. Quinteiro, & A. Gonçalves (Eds.), *ICT as Innovator Between Tourism and Culture* (pp. 213–229). IGI Global. https://doi.org/10.4018/978-1-7998-8165-0.ch013

Kinnunen, S., Ylä-Kujala, A., Marttonen-Arola, S., Kärri, T., & Baglee, D. (2018). Internet of Things in Asset Management: Insights from Industrial Professionals and Academia. *International Journal of Service Science, Management, Engineering, and Technology*, *9*(2), 104–119. doi:10.4018/IJSSMET.2018040105

Klein, A. Z., Sabino de Freitas, A., Machado, L., Freitas, J. C. Jr, Graziola, P. G. Jr, & Schlemmer, E. (2017). Virtual Worlds Applications for Management Education. In L. Tomei (Ed.), *Exploring the New Era of Technology-Infused Education* (pp. 279–299). Hershey, PA: IGI Global. doi:10.4018/978-1-5225-1709-2.ch017

Kővári, E., Saleh, M., & Steinbachné Hajmásy, G. (2022). The Impact of Corporate Digital Responsibility (CDR) on Internal Stakeholders' Satisfaction in Hungarian Upscale Hotels. In M. Valeri (Ed.), *New Governance and Management in Touristic Destinations* (pp. 35–51). IGI Global. https://doi.org/10.4018/978-1-6684-3889-3.ch003

Kożuch, B., & Jabłoński, A. (2017). Adopting the Concept of Business Models in Public Management. In M. Lewandowski & B. Kożuch (Eds.), *Public Sector Entrepreneurship and the Integration of Innovative Business Models* (pp. 10–46). Hershey, PA: IGI Global. doi:10.4018/978-1-5225-2215-7.ch002

Kumar, J., Adhikary, A., & Jha, A. (2017). Small Active Investors' Perceptions and Preferences Towards Tax Saving Mutual Fund Schemes in Eastern India: An Empirical Note. *International Journal of Asian Business and Information Management*, *8*(2), 35–45. doi:10.4018/IJABIM.2017040103

Latusi, S., & Fissore, M. (2021). Pilgrimage Routes to Happiness: Comparing the Camino de Santiago and Via Francigena. In A. Perinotto, V. Mayer, & J. Soares (Eds.), *Rebuilding and Restructuring the Tourism Industry: Infusion of Happiness and Quality of Life* (pp. 157–182). IGI Global. https://doi.org/10.4018/978-1-7998-7239-9.ch008

Lavassani, K. M., & Movahedi, B. (2017). Applications Driven Information Systems: Beyond Networks toward Business Ecosystems. *International Journal of Innovation in the Digital Economy*, *8*(1), 61–75. doi:10.4018/IJIDE.2017010104

Lazzareschi, V. H., & Brito, M. S. (2017). Strategic Information Management: Proposal of Business Project Model. In G. Jamil, A. Soares, & C. Pessoa (Eds.), *Handbook of Research on Information Management for Effective Logistics and Supply Chains* (pp. 59–88). Hershey, PA: IGI Global. doi:10.4018/978-1-5225-0973-8.ch004

Lechuga Sancho, M. P., & Martín Navarro, A. (2022). Evolution of the Literature on Social Responsibility in the Tourism Sector: A Systematic Literature Review. In G. Fernandes (Ed.), *Challenges and New Opportunities for Tourism in Inland Territories: Ecocultural Resources and Sustainable Initiatives* (pp. 169–186). IGI Global. https://doi.org/10.4018/978-1-7998-7339-6.ch010

Lederer, M., Kurz, M., & Lazarov, P. (2017). Usage and Suitability of Methods for Strategic Business Process Initiatives: A Multi Case Study Research. *International Journal of Productivity Management and Assessment Technologies*, *5*(1), 40–51. doi:10.4018/IJPMAT.2017010103

Lee, I. (2017). A Social Enterprise Business Model and a Case Study of Pacific Community Ventures (PCV). In V. Potocan, M. Üngan, & Z. Nedelko (Eds.), *Handbook of Research on Managerial Solutions in Non-Profit Organizations* (pp. 182–204). Hershey, PA: IGI Global. doi:10.4018/978-1-5225-0731-4.ch009

Leon, L. A., Seal, K. C., Przasnyski, Z. H., & Wiedenman, I. (2017). Skills and Competencies Required for Jobs in Business Analytics: A Content Analysis of Job Advertisements Using Text Mining. *International Journal of Business Intelligence Research*, *8*(1), 1–25. doi:10.4018/IJBIR.2017010101

Levy, C. L., & Elias, N. I. (2017). SOHO Users' Perceptions of Reliability and Continuity of Cloud-Based Services. In M. Moore (Ed.), *Cybersecurity Breaches and Issues Surrounding Online Threat Protection* (pp. 248–287). Hershey, PA: IGI Global. doi:10.4018/978-1-5225-1941-6.ch011

Levy, M. (2018). Change Management Serving Knowledge Management and Organizational Development: Reflections and Review. In N. Baporikar (Ed.), *Global Practices in Knowledge Management for Societal and Organizational Development* (pp. 256–270). Hershey, PA: IGI Global. doi:10.4018/978-1-5225-3009-1.ch012

Lewandowski, M. (2017). Public Organizations and Business Model Innovation: The Role of Public Service Design. In M. Lewandowski & B. Kożuch (Eds.), *Public Sector Entrepreneurship and the Integration of Innovative Business Models* (pp. 47–72). Hershey, PA: IGI Global. doi:10.4018/978-1-5225-2215-7.ch003

Lhannaoui, H., Kabbaj, M. I., & Bakkoury, Z. (2017). A Survey of Risk-Aware Business Process Modelling. *International Journal of Risk and Contingency Management, 6*(3), 14–26. doi:10.4018/IJRCM.2017070102

Li, J., Sun, W., Jiang, W., Yang, H., & Zhang, L. (2017). How the Nature of Exogenous Shocks and Crises Impact Company Performance?: The Effects of Industry Characteristics. *International Journal of Risk and Contingency Management, 6*(4), 40–55. doi:10.4018/IJRCM.2017100103

Lopez-Fernandez, M., Perez-Perez, M., Serrano-Bedia, A., & Cobo-Gonzalez, A. (2021). Small and Medium Tourism Enterprise Survival in Times of Crisis: "El Capricho de Gaudí. In D. Toubes & N. Araújo-Vila (Eds.), *Risk, Crisis, and Disaster Management in Small and Medium-Sized Tourism Enterprises* (pp. 103–129). IGI Global. doi:10.4018/978-1-7998-6996-2.ch005

Mahajan, A., Maidullah, S., & Hossain, M. R. (2022). Experience Toward Smart Tour Guide Apps in Travelling: An Analysis of Users' Reviews on Audio Odigos and Trip My Way. In R. Augusto Costa, F. Brandão, Z. Breda, & C. Costa (Eds.), *Planning and Managing the Experience Economy in Tourism* (pp. 255-273). IGI Global. https://doi.org/10.4018/978-1-7998-8775-1.ch014

Malega, P. (2017). Small and Medium Enterprises in the Slovak Republic: Status and Competitiveness of SMEs in the Global Markets and Possibilities of Optimization. In M. Vemić (Ed.), *Optimal Management Strategies in Small and Medium Enterprises* (pp. 102–124). Hershey, PA: IGI Global. doi:10.4018/978-1-5225-1949-2.ch006

Malewska, K. M. (2017). Intuition in Decision-Making on the Example of a Non-Profit Organization. In V. Potocan, M. Ünğan, & Z. Nedelko (Eds.), *Handbook of Research on Managerial Solutions in Non-Profit Organizations* (pp. 378–399). Hershey, PA: IGI Global. doi:10.4018/978-1-5225-0731-4.ch018

Maroofi, F. (2017). Entrepreneurial Orientation and Organizational Learning Ability Analysis for Innovation and Firm Performance. In N. Baporikar (Ed.), *Innovation and Shifting Perspectives in Management Education* (pp. 144–165). Hershey, PA: IGI Global. doi:10.4018/978-1-5225-1019-2.ch007

Marques, M., Moleiro, D., Brito, T. M., & Marques, T. (2021). Customer Relationship Management as an Important Relationship Marketing Tool: The Case of the Hospitality Industry in Estoril Coast. In M. Dinis, L. Bonixe, S. Lamy, & Z. Breda (Eds.), Impact of New Media in Tourism (pp. 39-56). IGI Global. https://doi.org/ doi:10.4018/978-1-7998-7095-1.ch003

Martins, P. V., & Zacarias, M. (2017). A Web-based Tool for Business Process Improvement. *International Journal of Web Portals*, *9*(2), 68–84. doi:10.4018/IJWP.2017070104

Matthies, B., & Coners, A. (2017). Exploring the Conceptual Nature of e-Business Projects. *Journal of Electronic Commerce in Organizations*, *15*(3), 33–63. doi:10.4018/JECO.2017070103

Mayer, V. F., Fraga, C. C., & Silva, L. C. (2021). Contributions of Neurosciences to Studies of Well-Being in Tourism. In A. Perinotto, V. Mayer, & J. Soares (Eds.), *Rebuilding and Restructuring the Tourism Industry: Infusion of Happiness and Quality of Life* (pp. 108–128). IGI Global. https://doi.org/10.4018/978-1-7998-7239-9.ch006

McKee, J. (2018). Architecture as a Tool to Solve Business Planning Problems. In M. Khosrow-Pour, D.B.A. (Ed.), Encyclopedia of Information Science and Technology, Fourth Edition (pp. 573-586). Hershey, PA: IGI Global. doi:10.4018/978-1-5225-2255-3.ch050

McMurray, A. J., Cross, J., & Caponecchia, C. (2018). The Risk Management Profession in Australia: Business Continuity Plan Practices. In N. Bajgoric (Ed.), *Always-On Enterprise Information Systems for Modern Organizations* (pp. 112–129). Hershey, PA: IGI Global. doi:10.4018/978-1-5225-3704-5.ch006

Meddah, I. H., & Belkadi, K. (2018). Mining Patterns Using Business Process Management. In R. Hamou (Ed.), *Handbook of Research on Biomimicry in Information Retrieval and Knowledge Management* (pp. 78–89). Hershey, PA: IGI Global. doi:10.4018/978-1-5225-3004-6.ch005

Melian, A. G., & Camprubí, R. (2021). The Accessibility of Museum Websites: The Case of Barcelona. In C. Eusébio, L. Teixeira, & M. Carneiro (Eds.), *ICT Tools and Applications for Accessible Tourism* (pp. 234–255). IGI Global. https://doi.org/10.4018/978-1-7998-6428-8.ch011

Mendes, L. (2017). TQM and Knowledge Management: An Integrated Approach Towards Tacit Knowledge Management. In D. Jaziri-Bouagina & G. Jamil (Eds.), *Handbook of Research on Tacit Knowledge Management for Organizational Success* (pp. 236–263). Hershey, PA: IGI Global. doi:10.4018/978-1-5225-2394-9.ch009

Menezes, V. D., & Cavagnaro, E. (2021). Communicating Sustainable Initiatives in the Hotel Industry: The Case of the Hotel Jakarta Amsterdam. In F. Brandão, Z. Breda, R. Costa, & C. Costa (Eds.), *Handbook of Research on the Role of Tourism in Achieving Sustainable Development Goals* (pp. 224-234). IGI Global. https://doi.org/10.4018/978-1-7998-5691-7.ch013

Menezes, V. D., & Cavagnaro, E. (2021). Communicating Sustainable Initiatives in the Hotel Industry: The Case of the Hotel Jakarta Amsterdam. In F. Brandão, Z. Breda, R. Costa, & C. Costa (Eds.), *Handbook of Research on the Role of Tourism in Achieving Sustainable Development Goals* (pp. 224-234). IGI Global. https://doi.org/10.4018/978-1-7998-5691-7.ch013

Mitas, O., Bastiaansen, M., & Boode, W. (2022). If You're Happy, I'm Happy: Emotion Contagion at a Tourist Information Center. In R. Augusto Costa, F. Brandão, Z. Breda, & C. Costa (Eds.), *Planning and Managing the Experience Economy in Tourism* (pp. 122-140). IGI Global. https://doi.org/10.4018/978-1-7998-8775-1.ch007

Mnjama, N. M. (2017). Preservation of Recorded Information in Public and Private Sector Organizations. In P. Jain & N. Mnjama (Eds.), *Managing Knowledge Resources and Records in Modern Organizations* (pp. 149–167). Hershey, PA: IGI Global. doi:10.4018/978-1-5225-1965-2.ch009

Mokoqama, M., & Fields, Z. (2017). Principles of Responsible Management Education (PRME): Call for Responsible Management Education. In Z. Fields (Ed.), *Collective Creativity for Responsible and Sustainable Business Practice* (pp. 229–241). Hershey, PA: IGI Global. doi:10.4018/978-1-5225-1823-5.ch012

Monteiro, A., Lopes, S., & Carbone, F. (2021). Academic Mobility: Bridging Tourism and Peace Education. In J. da Silva, Z. Breda, & F. Carbone (Eds.), *Role and Impact of Tourism in Peacebuilding and Conflict Transformation* (pp. 275-301). IGI Global. https://doi.org/10.4018/978-1-7998-5053-3.ch016

Muniapan, B. (2017). Philosophy and Management: The Relevance of Vedanta in Management. In P. Ordóñez de Pablos (Ed.), *Managerial Strategies and Solutions for Business Success in Asia* (pp. 124–139). Hershey, PA: IGI Global. doi:10.4018/978-1-5225-1886-0.ch007

Murad, S. E., & Dowaji, S. (2017). Using Value-Based Approach for Managing Cloud-Based Services. In A. Turuk, B. Sahoo, & S. Addya (Eds.), *Resource Management and Efficiency in Cloud Computing Environments* (pp. 33–60). Hershey, PA: IGI Global. doi:10.4018/978-1-5225-1721-4.ch002

Mutahar, A. M., Daud, N. M., Thurasamy, R., Isaac, O., & Abdulsalam, R. (2018). The Mediating of Perceived Usefulness and Perceived Ease of Use: The Case of Mobile Banking in Yemen. *International Journal of Technology Diffusion*, *9*(2), 21–40. doi:10.4018/IJTD.2018040102

Naidoo, V. (2017). E-Learning and Management Education at African Universities. In N. Baporikar (Ed.), *Management Education for Global Leadership* (pp. 181–201). Hershey, PA: IGI Global. doi:10.4018/978-1-5225-1013-0.ch009

Naidoo, V., & Igbinakhase, I. (2018). Opportunities and Challenges of Knowledge Retention in SMEs. In N. Baporikar (Ed.), *Knowledge Integration Strategies for Entrepreneurship and Sustainability* (pp. 70–94). Hershey, PA: IGI Global. doi:10.4018/978-1-5225-5115-7.ch004

Naumov, N., & Costandachi, G. (2021). Creativity and Entrepreneurship: Gastronomic Tourism in Mexico. In J. Soares (Ed.), *Innovation and Entrepreneurial Opportunities in Community Tourism* (pp. 90–108). IGI Global. https://doi.org/10.4018/978-1-7998-4855-4.ch006

Nayak, S., & Prabhu, N. (2017). Paradigm Shift in Management Education: Need for a Cross Functional Perspective. In N. Baporikar (Ed.), *Management Education for Global Leadership* (pp. 241–255). Hershey, PA: IGI Global. doi:10.4018/978-1-5225-1013-0.ch012

Nedelko, Z., & Potocan, V. (2017). Management Solutions in Non-Profit Organizations: Case of Slovenia. In V. Potocan, M. Üngan, & Z. Nedelko (Eds.), *Handbook of Research on Managerial Solutions in Non-Profit Organizations* (pp. 1–22). Hershey, PA: IGI Global. doi:10.4018/978-1-5225-0731-4.ch001

Nedelko, Z., & Potocan, V. (2017). Priority of Management Tools Utilization among Managers: International Comparison. In V. Wang (Ed.), *Encyclopedia of Strategic Leadership and Management* (pp. 1083–1094). Hershey, PA: IGI Global. doi:10.4018/978-1-5225-1049-9.ch075

Nedelko, Z., Raudeliūnienė, J., & Črešnar, R. (2018). Knowledge Dynamics in Supply Chain Management. In N. Baporikar (Ed.), *Knowledge Integration Strategies for Entrepreneurship and Sustainability* (pp. 150–166). Hershey, PA: IGI Global. doi:10.4018/978-1-5225-5115-7.ch008

Nguyen, H. T., & Hipsher, S. A. (2018). Innovation and Creativity Used by Private Sector Firms in a Resources-Constrained Environment. In S. Hipsher (Ed.), *Examining the Private Sector's Role in Wealth Creation and Poverty Reduction* (pp. 219–238). Hershey, PA: IGI Global. doi:10.4018/978-1-5225-3117-3.ch010

Obicci, P. A. (2017). Risk Sharing in a Partnership. In *Risk Management Strategies in Public-Private Partnerships* (pp. 115–152). Hershey, PA: IGI Global. doi:10.4018/978-1-5225-2503-5.ch004

Obidallah, W. J., & Raahemi, B. (2017). Managing Changes in Service Oriented Virtual Organizations: A Structural and Procedural Framework to Facilitate the Process of Change. *Journal of Electronic Commerce in Organizations, 15*(1), 59–83. doi:10.4018/JECO.2017010104

Ojo, O. (2017). Impact of Innovation on the Entrepreneurial Success in Selected Business Enterprises in South-West Nigeria. *International Journal of Innovation in the Digital Economy, 8*(2), 29–38. doi:10.4018/IJIDE.2017040103

Okdinawati, L., Simatupang, T. M., & Sunitiyoso, Y. (2017). Multi-Agent Reinforcement Learning for Value Co-Creation of Collaborative Transportation Management (CTM). *International Journal of Information Systems and Supply Chain Management, 10*(3), 84–95. doi:10.4018/IJISSCM.2017070105

Olivera, V. A., & Carrillo, I. M. (2021). Organizational Culture: A Key Element for the Development of Mexican Micro and Small Tourist Companies. In J. Soares (Ed.), *Innovation and Entrepreneurial Opportunities in Community Tourism* (pp. 227–242). IGI Global. doi:10.4018/978-1-7998-4855-4.ch013

Ossorio, M. (2022). Corporate Museum Experiences in Enogastronomic Tourism. In R. Augusto Costa, F. Brandão, Z. Breda, & C. Costa (Eds.), Planning and Managing the Experience Economy in Tourism (pp. 107-121). IGI Global. https://doi.org/ doi:10.4018/978-1-7998-8775-1.ch006

Ossorio, M. (2022). Enogastronomic Tourism in Times of Pandemic. In G. Fernandes (Ed.), *Challenges and New Opportunities for Tourism in Inland Territories: Ecocultural Resources and Sustainable Initiatives* (pp. 241–255). IGI Global. https://doi.org/10.4018/978-1-7998-7339-6.ch014

Özekici, Y. K. (2022). ICT as an Acculturative Agent and Its Role in the Tourism Context: Introduction, Acculturation Theory, Progress of the Acculturation Theory in Extant Literature. In C. Ramos, S. Quinteiro, & A. Gonçalves (Eds.), *ICT as Innovator Between Tourism and Culture* (pp. 42–66). IGI Global. https://doi.org/10.4018/978-1-7998-8165-0.ch004

Pal, K. (2018). Building High Quality Big Data-Based Applications in Supply Chains. In A. Kumar & S. Saurav (Eds.), *Supply Chain Management Strategies and Risk Assessment in Retail Environments* (pp. 1–24). Hershey, PA: IGI Global. doi:10.4018/978-1-5225-3056-5.ch001

Palos-Sanchez, P. R., & Correia, M. B. (2018). Perspectives of the Adoption of Cloud Computing in the Tourism Sector. In J. Rodrigues, C. Ramos, P. Cardoso, & C. Henriques (Eds.), *Handbook of Research on Technological Developments for Cultural Heritage and eTourism Applications* (pp. 377–400). Hershey, PA: IGI Global. doi:10.4018/978-1-5225-2927-9.ch018

Papadopoulou, G. (2021). Promoting Gender Equality and Women Empowerment in the Tourism Sector. In F. Brandão, Z. Breda, R. Costa, & C. Costa (Eds.), Handbook of Research on the Role of Tourism in Achieving Sustainable Development Goals (pp. 152-174). IGI Global. https://doi.org/ doi:10.4018/978-1-7998-5691-7.ch009

Papp-Váry, Á. F., & Tóth, T. Z. (2022). Analysis of Budapest as a Film Tourism Destination. In R. Baleiro & R. Pereira (Eds.), *Global Perspectives on Literary Tourism and Film-Induced Tourism* (pp. 257-279). IGI Global. https://doi.org/10.4018/978-1-7998-8262-6.ch014

Patiño, B. E. (2017). New Generation Management by Convergence and Individual Identity: A Systemic and Human-Oriented Approach. In N. Baporikar (Ed.), *Innovation and Shifting Perspectives in Management Education* (pp. 119–143). Hershey, PA: IGI Global. doi:10.4018/978-1-5225-1019-2.ch006

Patro, C. S. (2021). Digital Tourism: Influence of E-Marketing Technology. In M. Dinis, L. Bonixe, S. Lamy, & Z. Breda (Eds.), *Impact of New Media in Tourism* (pp. 234-254). IGI Global. https://doi.org/10.4018/978-1-7998-7095-1.ch014

Pawliczek, A., & Rössler, M. (2017). Knowledge of Management Tools and Systems in SMEs: Knowledge Transfer in Management. In A. Bencsik (Ed.), *Knowledge Management Initiatives and Strategies in Small and Medium Enterprises* (pp. 180–203). Hershey, PA: IGI Global. doi:10.4018/978-1-5225-1642-2.ch009

Pejic-Bach, M., Omazic, M. A., Aleksic, A., & Zoroja, J. (2018). Knowledge-Based Decision Making: A Multi-Case Analysis. In R. Leon (Ed.), *Managerial Strategies for Business Sustainability During Turbulent Times* (pp. 160–184). Hershey, PA: IGI Global. doi:10.4018/978-1-5225-2716-9.ch009

Perano, M., Hysa, X., & Calabrese, M. (2018). Strategic Planning, Cultural Context, and Business Continuity Management: Business Cases in the City of Shkoder. In A. Presenza & L. Sheehan (Eds.), *Geopolitics and Strategic Management in the Global Economy* (pp. 57–77). Hershey, PA: IGI Global. doi:10.4018/978-1-5225-2673-5.ch004

Pereira, R., Mira da Silva, M., & Lapão, L. V. (2017). IT Governance Maturity Patterns in Portuguese Healthcare. In S. De Haes & W. Van Grembergen (Eds.), *Strategic IT Governance and Alignment in Business Settings* (pp. 24–52). Hershey, PA: IGI Global. doi:10.4018/978-1-5225-0861-8.ch002

Pérez-Uribe, R. I., Torres, D. A., Jurado, S. P., & Prada, D. M. (2018). Cloud Tools for the Development of Project Management in SMEs. In R. Perez-Uribe, C. Salcedo-Perez, & D. Ocampo-Guzman (Eds.), *Handbook of Research on Intrapreneurship and Organizational Sustainability in SMEs* (pp. 95–120). Hershey, PA: IGI Global. doi:10.4018/978-1-5225-3543-0.ch005

Petrisor, I., & Cozmiuc, D. (2017). Global Supply Chain Management Organization at Siemens in the Advent of Industry 4.0. In L. Saglietto & C. Cezanne (Eds.), *Global Intermediation and Logistics Service Providers* (pp. 123–142). Hershey, PA: IGI Global. doi:10.4018/978-1-5225-2133-4.ch007

Pierce, J. M., Velliaris, D. M., & Edwards, J. (2017). A Living Case Study: A Journey Not a Destination. In N. Silton (Ed.), *Exploring the Benefits of Creativity in Education, Media, and the Arts* (pp. 158–178). Hershey, PA: IGI Global. doi:10.4018/978-1-5225-0504-4.ch008

Pipia, S., & Pipia, S. (2021). Challenges of Religious Tourism in the Conflict Region: An Example of Jerusalem. In E. Alaverdov & M. Bari (Eds.), *Global Development of Religious Tourism* (pp. 135-148). IGI Global. https://doi.org/10.4018/978-1-7998-5792-1.ch009

Poulaki, P., Kritikos, A., Vasilakis, N., & Valeri, M. (2022). The Contribution of Female Creativity to the Development of Gastronomic Tourism in Greece: The Case of the Island of Naxos in the South Aegean Region. In M. Valeri (Ed.), *New Governance and Management in Touristic Destinations* (pp. 246–258). IGI Global. https://doi.org/10.4018/978-1-6684-3889-3.ch015

Radosavljevic, M., & Andjelkovic, A. (2017). Multi-Criteria Decision Making Approach for Choosing Business Process for the Improvement: Upgrading of the Six Sigma Methodology. In J. Stanković, P. Delias, S. Marinković, & S. Rochhia (Eds.), *Tools and Techniques for Economic Decision Analysis* (pp. 225–247). Hershey, PA: IGI Global. doi:10.4018/978-1-5225-0959-2.ch011

Radovic, V. M. (2017). Corporate Sustainability and Responsibility and Disaster Risk Reduction: A Serbian Overview. In M. Camilleri (Ed.), *CSR 2.0 and the New Era of Corporate Citizenship* (pp. 147–164). Hershey, PA: IGI Global. doi:10.4018/978-1-5225-1842-6.ch008

Raghunath, K. M., Devi, S. L., & Patro, C. S. (2018). Impact of Risk Assessment Models on Risk Factors: A Holistic Outlook. In K. Strang, M. Korstanje, & N. Vajjhala (Eds.), *Research, Practices, and Innovations in Global Risk and Contingency Management* (pp. 134–153). Hershey, PA: IGI Global. doi:10.4018/978-1-5225-4754-9.ch008

Raman, A., & Goyal, D. P. (2017). Extending IMPLEMENT Framework for Enterprise Information Systems Implementation to Information System Innovation. In M. Tavana (Ed.), *Enterprise Information Systems and the Digitalization of Business Functions* (pp. 137–177). Hershey, PA: IGI Global. doi:10.4018/978-1-5225-2382-6.ch007

Rao, Y., & Zhang, Y. (2017). The Construction and Development of Academic Library Digital Special Subject Databases. In L. Ruan, Q. Zhu, & Y. Ye (Eds.), *Academic Library Development and Administration in China* (pp. 163–183). Hershey, PA: IGI Global. doi:10.4018/978-1-5225-0550-1.ch010

Ravasan, A. Z., Mohammadi, M. M., & Hamidi, H. (2018). An Investigation Into the Critical Success Factors of Implementing Information Technology Service Management Frameworks. In K. Jakobs (Ed.), *Corporate and Global Standardization Initiatives in Contemporary Society* (pp. 200–218). Hershey, PA: IGI Global. doi:10.4018/978-1-5225-5320-5.ch009

Rezaie, S., Mirabedini, S. J., & Abtahi, A. (2018). Designing a Model for Implementation of Business Intelligence in the Banking Industry. *International Journal of Enterprise Information Systems*, *14*(1), 77–103. doi:10.4018/IJEIS.2018010105

Richards, V., Matthews, N., Williams, O. J., & Khan, Z. (2021). The Challenges of Accessible Tourism Information Systems for Tourists With Vision Impairment: Sensory Communications Beyond the Screen. In C. Eusébio, L. Teixeira, & M. Carneiro (Eds.), *ICT Tools and Applications for Accessible Tourism* (pp. 26–54). IGI Global. https://doi.org/10.4018/978-1-7998-6428-8.ch002

Rodrigues de Souza Neto, V., & Marques, O. (2021). Rural Tourism Fostering Welfare Through Sustainable Development: A Conceptual Approach. In A. Perinotto, V. Mayer, & J. Soares (Eds.), *Rebuilding and Restructuring the Tourism Industry: Infusion of Happiness and Quality of Life* (pp. 38–57). IGI Global. https://doi.org/10.4018/978-1-7998-7239-9.ch003

Romano, L., Grimaldi, R., & Colasuonno, F. S. (2017). Demand Management as a Success Factor in Project Portfolio Management. In L. Romano (Ed.), *Project Portfolio Management Strategies for Effective Organizational Operations* (pp. 202–219). Hershey, PA: IGI Global. doi:10.4018/978-1-5225-2151-8.ch008

Rubio-Escuderos, L., & García-Andreu, H. (2021). Competitiveness Factors of Accessible Tourism E-Travel Agencies. In C. Eusébio, L. Teixeira, & M. Carneiro (Eds.), *ICT Tools and Applications for Accessible Tourism* (pp. 196–217). IGI Global. https://doi.org/10.4018/978-1-7998-6428-8.ch009

Rucci, A. C., Porto, N., Darcy, S., & Becka, L. (2021). Smart and Accessible Cities?: Not Always – The Case for Accessible Tourism Initiatives in Buenos Aries and Sydney. In C. Eusébio, L. Teixeira, & M. Carneiro (Eds.), *ICT Tools and Applications for Accessible Tourism* (pp. 115–145). IGI Global. https://doi.org/10.4018/978-1-7998-6428-8.ch006

Ruhi, U. (2018). Towards an Interdisciplinary Socio-Technical Definition of Virtual Communities. In M. Khosrow-Pour, D.B.A. (Ed.), Encyclopedia of Information Science and Technology, Fourth Edition (pp. 4278-4295). Hershey, PA: IGI Global. doi:10.4018/978-1-5225-2255-3.ch371

Ryan, L., Catena, M., Ros, P., & Stephens, S. (2021). Designing Entrepreneurial Ecosystems to Support Resource Management in the Tourism Industry. In V. Costa, A. Moura, & M. Mira (Eds.), *Handbook of Research on Human Capital and People Management in the Tourism Industry* (pp. 265–281). IGI Global. https://doi.org/10.4018/978-1-7998-4318-4.ch013

Sabuncu, I. (2021). Understanding Tourist Perceptions and Expectations During Pandemic Through Social Media Big Data. In M. Demir, A. Dalgıç, & F. Ergen (Eds.), *Handbook of Research on the Impacts and Implications of COVID-19 on the Tourism Industry* (pp. 330–350). IGI Global. https://doi.org/10.4018/978-1-7998-8231-2.ch016

Safari, M. R., & Jiang, Q. (2018). The Theory and Practice of IT Governance Maturity and Strategies Alignment: Evidence From Banking Industry. *Journal of Global Information Management, 26*(2), 127–146. doi:10.4018/JGIM.2018040106

Sahoo, J., Pati, B., & Mohanty, B. (2017). Knowledge Management as an Academic Discipline: An Assessment. In B. Gunjal (Ed.), *Managing Knowledge and Scholarly Assets in Academic Libraries* (pp. 99–126). Hershey, PA: IGI Global. doi:10.4018/978-1-5225-1741-2.ch005

Saini, D. (2017). Relevance of Teaching Values and Ethics in Management Education. In N. Baporikar (Ed.), *Management Education for Global Leadership* (pp. 90–111). Hershey, PA: IGI Global. doi:10.4018/978-1-5225-1013-0.ch005

Sambhanthan, A. (2017). Assessing and Benchmarking Sustainability in Organisations: An Integrated Conceptual Model. *International Journal of Systems and Service-Oriented Engineering*, *7*(4), 22–43. doi:10.4018/IJSSOE.2017100102

Sambhanthan, A., & Potdar, V. (2017). A Study of the Parameters Impacting Sustainability in Information Technology Organizations. *International Journal of Knowledge-Based Organizations*, *7*(3), 27–39. doi:10.4018/IJKBO.2017070103

Sánchez-Fernández, M. D., & Manríquez, M. R. (2018). The Entrepreneurial Spirit Based on Social Values: The Digital Generation. In P. Isaias & L. Carvalho (Eds.), *User Innovation and the Entrepreneurship Phenomenon in the Digital Economy* (pp. 173–193). Hershey, PA: IGI Global. doi:10.4018/978-1-5225-2826-5.ch009

Sanchez-Ruiz, L., & Blanco, B. (2017). Process Management for SMEs: Barriers, Enablers, and Benefits. In M. Vemić (Ed.), *Optimal Management Strategies in Small and Medium Enterprises* (pp. 293–319). Hershey, PA: IGI Global. doi:10.4018/978-1-5225-1949-2.ch014

Sanz, L. F., Gómez-Pérez, J., & Castillo-Martinez, A. (2018). Analysis of the European ICT Competence Frameworks. In V. Ahuja & S. Rathore (Eds.), *Multidisciplinary Perspectives on Human Capital and Information Technology Professionals* (pp. 225–245). Hershey, PA: IGI Global. doi:10.4018/978-1-5225-5297-0.ch012

Sarvepalli, A., & Godin, J. (2017). Business Process Management in the Classroom. *Journal of Cases on Information Technology*, *19*(2), 17–28. doi:10.4018/JCIT.2017040102

Saxena, G. G., & Saxena, A. (2021). Host Community Role in Medical Tourism Development. In M. Singh & S. Kumaran (Eds.), *Growth of the Medical Tourism Industry and Its Impact on Society: Emerging Research and Opportunities* (pp. 105–127). IGI Global. https://doi.org/10.4018/978-1-7998-3427-4.ch006

Saygili, E. E., Ozturkoglu, Y., & Kocakulah, M. C. (2017). End Users' Perceptions of Critical Success Factors in ERP Applications. *International Journal of Enterprise Information Systems*, *13*(4), 58–75. doi:10.4018/IJEIS.2017100104

Saygili, E. E., & Saygili, A. T. (2017). Contemporary Issues in Enterprise Information Systems: A Critical Review of CSFs in ERP Implementations. In M. Tavana (Ed.), *Enterprise Information Systems and the Digitalization of Business Functions* (pp. 120–136). Hershey, PA: IGI Global. doi:10.4018/978-1-5225-2382-6.ch006

Schwaiger, K. M., & Zehrer, A. (2021). The COVID-19 Pandemic and Organizational Resilience in Hospitality Family Firms: A Qualitative Approach. In A. Zehrer, G. Glowka, K. Schwaiger, & V. Ranacher-Lackner (Eds.), *Resiliency Models and Addressing Future Risks for Family Firms in the Tourism Industry* (pp. 32–49). IGI Global. https://doi.org/10.4018/978-1-7998-7352-5.ch002

Scott, N., & Campos, A. C. (2022). Cognitive Science of Tourism Experiences. In R. Augusto Costa, F. Brandão, Z. Breda, & C. Costa (Eds.), Planning and Managing the Experience Economy in Tourism (pp. 1-21). IGI Global. https://doi.org/ doi:10.4018/978-1-7998-8775-1.ch001

Seidenstricker, S., & Antonino, A. (2018). Business Model Innovation-Oriented Technology Management for Emergent Technologies. In M. Khosrow-Pour, D.B.A. (Ed.), Encyclopedia of Information Science and Technology, Fourth Edition (pp. 4560-4569). Hershey, PA: IGI Global. doi:10.4018/978-1-5225-2255-3.ch396

Selvi, M. S. (2021). Changes in Tourism Sales and Marketing Post COVID-19. In M. Demir, A. Dalgıç, & F. Ergen (Eds.), *Handbook of Research on the Impacts and Implications of COVID-19 on the Tourism Industry* (pp. 437–460). IGI Global. doi:10.4018/978-1-7998-8231-2.ch021

Senaratne, S., & Gunarathne, A. D. (2017). Excellence Perspective for Management Education from a Global Accountants' Hub in Asia. In N. Baporikar (Ed.), *Management Education for Global Leadership* (pp. 158–180). Hershey, PA: IGI Global. doi:10.4018/978-1-5225-1013-0.ch008

Sensuse, D. I., & Cahyaningsih, E. (2018). Knowledge Management Models: A Summative Review. *International Journal of Information Systems in the Service Sector, 10*(1), 71–100. doi:10.4018/IJISSS.2018010105

Seth, M., Goyal, D., & Kiran, R. (2017). Diminution of Impediments in Implementation of Supply Chain Management Information System for Enhancing its Effectiveness in Indian Automobile Industry. *Journal of Global Information Management, 25*(3), 1–20. doi:10.4018/JGIM.2017070101

Seyal, A. H., & Rahman, M. N. (2017). Investigating Impact of Inter-Organizational Factors in Measuring ERP Systems Success: Bruneian Perspectives. In M. Tavana (Ed.), *Enterprise Information Systems and the Digitalization of Business Functions* (pp. 178–204). Hershey, PA: IGI Global. doi:10.4018/978-1-5225-2382-6.ch008

Shaqrah, A. A. (2018). Analyzing Business Intelligence Systems Based on 7s Model of McKinsey. *International Journal of Business Intelligence Research, 9*(1), 53–63. doi:10.4018/IJBIR.2018010104

Sharma, A. J. (2017). Enhancing Sustainability through Experiential Learning in Management Education. In N. Baporikar (Ed.), *Management Education for Global Leadership* (pp. 256–274). Hershey, PA: IGI Global. doi:10.4018/978-1-5225-1013-0.ch013

Shetty, K. P. (2017). Responsible Global Leadership: Ethical Challenges in Management Education. In N. Baporikar (Ed.), *Innovation and Shifting Perspectives in Management Education* (pp. 194–223). Hershey, PA: IGI Global. doi:10.4018/978-1-5225-1019-2.ch009

Sinthupundaja, J., & Kohda, Y. (2017). Effects of Corporate Social Responsibility and Creating Shared Value on Sustainability. *International Journal of Sustainable Entrepreneurship and Corporate Social Responsibility, 2*(1), 27–38. doi:10.4018/IJSECSR.2017010103

Škarica, I., & Hrgović, A. V. (2018). Implementation of Total Quality Management Principles in Public Health Institutes in the Republic of Croatia. *International Journal of Productivity Management and Assessment Technologies, 6*(1), 1–16. doi:10.4018/IJPMAT.2018010101

Skokic, V. (2021). How Small Hotel Owners Practice Resilience: Longitudinal Study Among Small Family Hotels in Croatia. In A. Zehrer, G. Glowka, K. Schwaiger, & V. Ranacher-Lackner (Eds.), *Resiliency Models and Addressing Future Risks for Family Firms in the Tourism Industry* (pp. 50–73). IGI Global. doi:10.4018/978-1-7998-7352-5.ch003

Smuts, H., Kotzé, P., Van der Merwe, A., & Loock, M. (2017). Framework for Managing Shared Knowledge in an Information Systems Outsourcing Context. *International Journal of Knowledge Management, 13*(4), 1–30. doi:10.4018/IJKM.2017100101

Sousa, M. J., Cruz, R., Dias, I., & Caracol, C. (2017). Information Management Systems in the Supply Chain. In G. Jamil, A. Soares, & C. Pessoa (Eds.), *Handbook of Research on Information Management for Effective Logistics and Supply Chains* (pp. 469–485). Hershey, PA: IGI Global. doi:10.4018/978-1-5225-0973-8.ch025

Spremic, M., Turulja, L., & Bajgoric, N. (2018). Two Approaches in Assessing Business Continuity Management Attitudes in the Organizational Context. In N. Bajgoric (Ed.), *Always-On Enterprise Information Systems for Modern Organizations* (pp. 159–183). Hershey, PA: IGI Global. doi:10.4018/978-1-5225-3704-5.ch008

Steenkamp, A. L. (2018). Some Insights in Computer Science and Information Technology. In *Examining the Changing Role of Supervision in Doctoral Research Projects: Emerging Research and Opportunities* (pp. 113–133). Hershey, PA: IGI Global. doi:10.4018/978-1-5225-2610-0.ch005

Stipanović, C., Rudan, E., & Zubović, V. (2022). Reaching the New Tourist Through Creativity: Sustainable Development Challenges in Croatian Coastal Towns. In M. Valeri (Ed.), *New Governance and Management in Touristic Destinations* (pp. 231–245). IGI Global. https://doi.org/10.4018/978-1-6684-3889-3.ch014

Tabach, A., & Croteau, A. (2017). Configurations of Information Technology Governance Practices and Business Unit Performance. *International Journal of IT/Business Alignment and Governance, 8*(2), 1–27. doi:10.4018/IJITBAG.2017070101

Talaue, G. M., & Iqbal, T. (2017). Assessment of e-Business Mode of Selected Private Universities in the Philippines and Pakistan. *International Journal of Online Marketing, 7*(4), 63–77. doi:10.4018/IJOM.2017100105

Tam, G. C. (2017). Project Manager Sustainability Competence. In *Managerial Strategies and Green Solutions for Project Sustainability* (pp. 178–207). Hershey, PA: IGI Global. doi:10.4018/978-1-5225-2371-0.ch008

Tambo, T. (2018). Fashion Retail Innovation: About Context, Antecedents, and Outcome in Technological Change Projects. In I. Management Association (Ed.), Fashion and Textiles: Breakthroughs in Research and Practice (pp. 233-260). Hershey, PA: IGI Global. https://doi.org/ doi:10.4018/978-1-5225-3432-7.ch010

Tantau, A. D., & Frăţilă, L. C. (2018). Information and Management System for Renewable Energy Business. In *Entrepreneurship and Business Development in the Renewable Energy Sector* (pp. 200–244). Hershey, PA: IGI Global. doi:10.4018/978-1-5225-3625-3.ch006

Teixeira, N., Pardal, P. N., & Rafael, B. G. (2018). Internationalization, Financial Performance, and Organizational Challenges: A Success Case in Portugal. In L. Carvalho (Ed.), *Handbook of Research on Entrepreneurial Ecosystems and Social Dynamics in a Globalized World* (pp. 379–423). Hershey, PA: IGI Global. doi:10.4018/978-1-5225-3525-6.ch017

Teixeira, P., Teixeira, L., Eusébio, C., Silva, S., & Teixeira, A. (2021). The Impact of ICTs on Accessible Tourism: Evidence Based on a Systematic Literature Review. In C. Eusébio, L. Teixeira, & M. Carneiro (Eds.), *ICT Tools and Applications for Accessible Tourism* (pp. 1–25). IGI Global. doi:10.4018/978-1-7998-6428-8.ch001

Trad, A., & Kalpić, D. (2018). The Business Transformation Framework, Agile Project and Change Management. In M. Khosrow-Pour, D.B.A. (Ed.), Encyclopedia of Information Science and Technology, Fourth Edition (pp. 620-635). Hershey, PA: IGI Global. https://doi.org/ doi:10.4018/978-1-5225-2255-3.ch054

Trad, A., & Kalpić, D. (2018). The Business Transformation and Enterprise Architecture Framework: The Financial Engineering E-Risk Management and E-Law Integration. In B. Sergi, F. Fidanoski, M. Ziolo, & V. Naumovski (Eds.), *Regaining Global Stability After the Financial Crisis* (pp. 46–65). Hershey, PA: IGI Global. doi:10.4018/978-1-5225-4026-7.ch003

Trengereid, V. (2022). Conditions of Network Engagement: The Quest for a Common Good. In R. Augusto Costa, F. Brandão, Z. Breda, & C. Costa (Eds.), *Planning and Managing the Experience Economy in Tourism* (pp. 69-84). IGI Global. https://doi.org/10.4018/978-1-7998-8775-1.ch004

Turulja, L., & Bajgoric, N. (2018). Business Continuity and Information Systems: A Systematic Literature Review. In N. Bajgoric (Ed.), *Always-On Enterprise Information Systems for Modern Organizations* (pp. 60–87). Hershey, PA: IGI Global. doi:10.4018/978-1-5225-3704-5.ch004

Vargas-Hernández, J. G. (2017). Professional Integrity in Business Management Education. In N. Baporikar (Ed.), *Management Education for Global Leadership* (pp. 70–89). Hershey, PA: IGI Global. doi:10.4018/978-1-5225-1013-0.ch004

Varnacı Uzun, F. (2021). The Destination Preferences of Foreign Tourists During the COVID-19 Pandemic and Attitudes Towards: Marmaris, Turkey. In M. Demir, A. Dalgıç, & F. Ergen (Eds.), *Handbook of Research on the Impacts and Implications of COVID-19 on the Tourism Industry* (pp. 285–306). IGI Global. https://doi.org/10.4018/978-1-7998-8231-2.ch014

Vasista, T. G., & AlAbdullatif, A. M. (2017). Role of Electronic Customer Relationship Management in Demand Chain Management: A Predictive Analytic Approach. *International Journal of Information Systems and Supply Chain Management*, *10*(1), 53–67. doi:10.4018/IJISSCM.2017010104

Vieru, D., & Bourdeau, S. (2017). Survival in the Digital Era: A Digital Competence-Based Multi-Case Study in the Canadian SME Clothing Industry. *International Journal of Social and Organizational Dynamics in IT*, *6*(1), 17–34. doi:10.4018/IJSODIT.2017010102

Vijayan, G., & Kamarulzaman, N. H. (2017). An Introduction to Sustainable Supply Chain Management and Business Implications. In M. Khan, M. Hussain, & M. Ajmal (Eds.), *Green Supply Chain Management for Sustainable Business Practice* (pp. 27–50). Hershey, PA: IGI Global. doi:10.4018/978-1-5225-0635-5.ch002

Vlachvei, A., & Notta, O. (2017). Firm Competitiveness: Theories, Evidence, and Measurement. In A. Vlachvei, O. Notta, K. Karantininis, & N. Tsounis (Eds.), *Factors Affecting Firm Competitiveness and Performance in the Modern Business World* (pp. 1–42). Hershey, PA: IGI Global. doi:10.4018/978-1-5225-0843-4.ch001

Wang, C., Schofield, M., Li, X., & Ou, X. (2017). Do Chinese Students in Public and Private Higher Education Institutes Perform at Different Level in One of the Leadership Skills: Critical Thinking?: An Exploratory Comparison. In V. Wang (Ed.), *Encyclopedia of Strategic Leadership and Management* (pp. 160–181). Hershey, PA: IGI Global. doi:10.4018/978-1-5225-1049-9.ch013

Wang, J. (2017). Multi-Agent based Production Management Decision System Modelling for the Textile Enterprise. *Journal of Global Information Management*, *25*(4), 1–15. doi:10.4018/JGIM.2017100101

Wiedemann, A., & Gewald, H. (2017). Examining Cross-Domain Alignment: The Correlation of Business Strategy, IT Management, and IT Business Value. *International Journal of IT/Business Alignment and Governance*, *8*(1), 17–31. doi:10.4018/IJITBAG.2017010102

Wolf, R., & Thiel, M. (2018). Advancing Global Business Ethics in China: Reducing Poverty Through Human and Social Welfare. In S. Hipsher (Ed.), *Examining the Private Sector's Role in Wealth Creation and Poverty Reduction* (pp. 67–84). Hershey, PA: IGI Global. doi:10.4018/978-1-5225-3117-3.ch004

Yablonsky, S. (2018). Innovation Platforms: Data and Analytics Platforms. In *Multi-Sided Platforms (MSPs) and Sharing Strategies in the Digital Economy: Emerging Research and Opportunities* (pp. 72–95). Hershey, PA: IGI Global. doi:10.4018/978-1-5225-5457-8.ch003

Yaşar, B. (2021). The Impact of COVID-19 on Volatility of Tourism Stocks: Evidence From BIST Tourism Index. In M. Demir, A. Dalgıç, & F. Ergen (Eds.), *Handbook of Research on the Impacts and Implications of COVID-19 on the Tourism Industry* (pp. 23–44). IGI Global. https://doi.org/10.4018/978-1-7998-8231-2.ch002

Yusoff, A., Ahmad, N. H., & Halim, H. A. (2017). Agropreneurship among Gen Y in Malaysia: The Role of Academic Institutions. In N. Ahmad, T. Ramayah, H. Halim, & S. Rahman (Eds.), *Handbook of Research on Small and Medium Enterprises in Developing Countries* (pp. 23–47). Hershey, PA: IGI Global. doi:10.4018/978-1-5225-2165-5.ch002

Zacher, D., & Pechlaner, H. (2021). Resilience as an Opportunity Approach: Challenges and Perspectives for Private Sector Participation on a Community Level. In A. Zehrer, G. Glowka, K. Schwaiger, & V. Ranacher-Lackner (Eds.), *Resiliency Models and Addressing Future Risks for Family Firms in the Tourism Industry* (pp. 75–102). IGI Global. https://doi.org/10.4018/978-1-7998-7352-5.ch004

Zanin, F., Comuzzi, E., & Costantini, A. (2018). The Effect of Business Strategy and Stock Market Listing on the Use of Risk Assessment Tools. In *Management Control Systems in Complex Settings: Emerging Research and Opportunities* (pp. 145–168). Hershey, PA: IGI Global. doi:10.4018/978-1-5225-3987-2.ch007

Zgheib, P. W. (2017). Corporate Innovation and Intrapreneurship in the Middle East. In P. Zgheib (Ed.), *Entrepreneurship and Business Innovation in the Middle East* (pp. 37–56). Hershey, PA: IGI Global. doi:10.4018/978-1-5225-2066-5.ch003

About the Author

Amiram Porath is an expert on innovation, with a specific focus on collaborative research and technology transfer. He has also been teaching entrepreneurship, in several programs over the years. since 2000 he has been active in the international arena regarding these topics and has consulted CREST and the EU commission. In 2008 Italy has awarded him the title "Cavaliere" for his contribution to the development of its scientific and industrial ties. Since 2012 Dr. Porath has been consulting public and governmental entities such as the OCS and SII.

Index

A

Advisory Board (AD) 22, 34

B

Beta-Site Testing 114
Bootstrap 37, 45, 48, 50, 52, 55

C

Complete Firm 1-2, 6, 13-14, 26-27, 30, 34, 37-38, 47, 49-50, 53-57, 59-63, 65-68, 70-72, 74, 79-81, 85-95, 105, 121-122, 125, 127, 130
customer satisfaction 119, 122

E

economic growth 5, 11, 14, 50, 65, 95, 100, 114
economic system 98
Ecosystem 3-6, 8-9, 11, 26-35, 38, 40, 46, 67-68, 77, 83, 99-100, 111, 113, 128
Environmental Fit 68

F

Financial Aid 98, 106, 114
Firm Growth 9-10, 61, 67-68, 95, 130

I

Israeli Innovation Authority 128, 131

M

managerial capabilities 21, 50, 55, 58, 66, 68, 70-71, 79, 85-86, 91, 94, 102, 106, 113, 122, 124-127
managerial differences 71
Market Entry 91, 107-110, 113-114
Market Failure 97-100, 113-115
military intelligence 117

N

Non-Financial Aid 106, 114
Non-Technical SU 37, 50, 52

O

Organizational Culture 2, 9, 17, 21, 47, 51, 60, 62, 67, 69, 72, 75-76, 89, 92, 122
Organizational Structure 16, 22, 45, 48, 53, 55, 57, 60, 69, 72, 75-77, 80, 119, 121

P

Perpetual/Repetitive Entrepreneur 83

S

Social Start-Up 38, 52
Start-Up (SU) 35
start-up company 5, 117
Start-Up firms 2

T

Technological Start-Up (TSU) 13, 35
Technology Innovation Incubators Program 131
Transient Capabilities 131

V

Value Creation Process 64, 71, 83

www.ingramcontent.com/pod-product-compliance
Lightning Source LLC
LaVergne TN
LVHW061533290926
849162LV00004B/781